THE MUSIC OF
LISZT

THE MUSIC OF

LISZT

Second Revised Edition

by

HUMPHREY SEARLE

DOVER PUBLICATIONS, INC.
NEW YORK

Published in Canada by General Publishing Company, Ltd., 30 Lesmill Road, Don Mills, Toronto,
Ontario.
Published in the United Kingdom by Constable
and Company, Ltd., 10 Orange Street, London WC 2.

This Dover edition, first published in 1966, is an
unabridged and revised republication of the work
first published by Williams & Norgate Ltd., London,
in 1954. It contains a new Preface by the author.

Library of Congress Catalog Card Number: 66-27581

Manufactured in the United States of America
Dover Publications, Inc.
180 Varick Street
New York, N.Y. 10014

In Memory of

CONSTANT LAMBERT

ACKNOWLEDGMENTS

MY THANKS are due to the President and Council of the Royal Musical Association for permission to reproduce those parts of Chapter IV which originally appeared in " Proceedings of the Royal Musical Association, Vol. LXXVIII " ; and to Messrs. Breitkopf & Härtel for permission to reprint the quotation from Busoni's edition of the Paganini-Liszt Étude No. 1.

H. S.

PREFACE TO THE SECOND EDITION

THE ISSUE of a second edition has made it possible to make some corrections and also to include discussion of some of Liszt's works which have only come to light since 1954; these include the *Historical Hungarian Portraits*, the fourth *Valse Oubliée*, and the *Bagatelle sans Tonalité*. In addition, the catalogue of works and bibliography have been brought up to date.

New York, 1966   H.S.

PREFACE TO THE FIRST EDITION

To WRITE a complete account of Liszt's music would need several volumes and a lifetime of work; and this book cannot attempt to do that. It only claims to give a general survey of Liszt's compositions, and in particular to draw attention to a number of his works which deserve more attention than they usually receive. I feel that it is useless to attempt to add to the biographies of Liszt which already exist—in particular the excellent one by Mr. Sacheverell Sitwell, which so admirably evokes the times in which Liszt lived—but a biographical summary is included which gives some idea of the relation in time between Liszt's life and his works. The works are discussed in chronological periods, and within each period are subdivided according to the medium they are written for; the normal order in each chapter is: piano music, other instrumental music, orchestral works, works for piano and orchestra, organ music, sacred choral works, secular choral works, songs and recitations. An alphabetical index of works at the end of the book gives the number of the page on which each work is referred to, and also the number under which it will be found in the catalogue. Naturally it has been impossible to discuss every one of the seven hundred works which Liszt wrote, and I must ask the indulgence of those who may complain that their favourites have been passed over in silence. All that I have tried to do is to give as much information as possible about all Liszt's more important and interesting works, and hope that this may stimulate others to pursue researches for themselves. For those who read German a great deal of useful information may be found in the two volumes of Peter Raabe's "Franz Liszt" (Stuttgart, 1931), a work to which any subsequent writer on Liszt must record his indebtedness; Raabe was for many years Director of the Liszt Museum at Weimar, and thus had access to a great many unpublished MSS. and letters. Though one may not always agree with Raabe's critical judgments, his book remains the main authority on the subject, pending the publication of the monumental researches of Professor Emile Haraszti. Further useful sources will be found listed in the bibliography.

London, 1954   H.S.

CONTENTS

THE MUSIC OF
LISZT

CHAPTER I

THE EARLY WORKS (1822-39)

LISZT WAS one of the most prolific of all the great composers. Although out of the seven hundred works or so which will be found in the catalogue at the end of this book many are new versions of earlier pieces, or transcriptions for a different medium of works by other composers or by Liszt himself, nevertheless, when one takes into account the enormous number of activities which he managed to crowd into his very varied life, his output remains astonishing. Of these works well over half are for the piano (including piano duet and two pianos) and in his early period, before he set off on his travels as a virtuoso pianist, he wrote for practically no other medium; even the works which he wrote for piano and orchestra at this time do not appear to have been scored by Liszt himself.

Liszt's earliest recorded composition (now lost) was a Tantum Ergo for choir, written in 1822, when he was an eleven-year-old pupil of Salieri in Vienna; in later years he dimly remembered this as being similar in mood to the Tantum Ergo of 1869. Apart from this work, and the short opera Don Sanche, all his youthful compositions were, naturally enough, for piano. The earliest surviving piece is a variation, also written in 1822, on the famous theme of Diabelli, for the collection to which fifty Austrian composers were invited to contribute one variation each, and to which Beethoven replied by sending thirty-three; other composers taking part included Schubert, Czerny, Hummel and Kalkbrenner, so this was a decided honour for Liszt. The variations were arranged in alphabetical order of the composers' names, and Liszt's is thus the twenty-fourth. It is published in the Breitkopf Collected Edition of Liszt's works; though certainly it is very competently written, it shows practically no individuality at all.

The same is true of the majority of the works written in his teens. Most of these were composed after his removal to Paris, and already in 1824 Czerny wrote to Adam Liszt that Franz had ready " 2 Rondos di bravura, for which offers have been made here, but I will not sell, 1 Rondo, 1 Fantasia, Variations on several themes, and 1 Amusement or rather Quodlibet on various themes of Rossini and Spontini, which he played to His Majesty with much applause." In the following year he mentions also two Concertos, a Sonata for four hands, a Trio and a Quintetto, remarking that the Concertos make those of Hummel seem quite easy by comparison.

Of all these works the only surviving ones* are the *Huit Variations* in A flat, dedicated to Sebastian Érard, the founder of the well-known piano and harp manufacturing firm, who helped Liszt considerably during this time; the *Variations brillantes sur un thème de G. Rossini*, a rare work of which copies are only to be found in the British Museum and the library of the Gesellschaft der Musikfreunde, Vienna (the theme is from Rossini's *Ermione*); the *Impromptu brillant sur des thèmes de Rossini et Spontini*, from the *Donna del Lago* and *Armida* of the former, and the *Olympia* and *Fernand Cortez* of the latter; the *Allegro di Bravura* and *Rondo di Bravura*, dedicated to Count Thaddeus Amadé, one of the Hungarian magnates who helped to provide for Liszt's education; the opera *Don Sanche ou le Château d'Amour*, which was produced at the Paris Opera on 17 October 1825; a Scherzo in G minor dating from 1827; two pieces called " Zum Andenken " (1828), which are interesting as being Liszt's first essays in the Hungarian style; and, most important of all, the *Étude en 48 Exercices dans tous les Tons Majeurs et Mineurs*—the earliest version of the Transcendental Studies. In addition the Liszt Museum at Weimar possess sketches for a piano concerto which is the forerunner of the so-called *Malediction* for piano and strings, and clearly dates from about this time.

The majority of these works need little discussion; they are the products of a clever schoolboy who happened also to be a

---

* The rather complex question of the opus numbers given to these early works has been satisfactorily dealt with by Busoni in his preface to the Collected Edition of the Études, Vol. I. This is republished in his *Von der Einheit der Musik* (English translation in preparation). Further details regarding Liszt's lost compositions can be found in Friedrich Schnapp: Verschollene Kompositionen Franz Liszts, published in " Von Deutscher Tonkunst " (Festschrift für Peter Raabe), Peters, Leipzig, 1942; this is a most detailed and valuable survey.

brilliant pianist and to possess a good deal of imagination. The general style is that of Liszt's master Czerny; the piano writing is always competent and often brilliant, but there is little in these pieces to indicate the experimental and revolutionary composer that Liszt was later to become. The most interesting musically (apart from the Studies, to which we shall return in connection with their later versions) is perhaps the little Scherzo in G minor, with its wide leaps and free use of the diminished seventh; the Hungarian pieces, of which the title " In Memory " may perhaps refer to the death of Liszt's father in the previous year, also deserve attention.

(The unpublished MS. is reported to be in the Prussian State Library in Berlin; perhaps publication will be possible in due course.)

The one-act opera, *Don Sanche, ou le Château d'Amour*, has a text by Théaulon and Rancé after a story by Claris de Florian (1755-94). The plot is a simple one. The Castle of Love can only be entered by those who both love and are loved. The knight Don Sanche is in love with Princess Elzire, but she wishes to marry the Prince of Navarre. However the master of the castle, the wizard Alidor, resolves to help Don Sanche. He leads the Princess and her followers astray in a storm and brings them near the castle, but will only allow her to enter if she loves Don Sanche. She refuses; but while she is asleep in the wood the savage Romuald (who is Alidor in disguise) comes to kidnap her. Don Sanche tries to defend her, and is carried in after a fight, apparently mortally wounded. Now at last her heart is touched and she confesses her love for Don Sanche; Alidor reveals himself, and the happy pair enter the castle, where a brilliant feast is prepared for them. The librettists saw to it that almost every theatrically effective device in the repertoire was included, among them peasant dances, Cupids descending from clouds, a Sleep aria, a storm scene, a fight off-stage, a funeral march and a final ballet. The text has been published by Roullet, Librairie de l'Académie royale de Musique; the music was long considered to be lost (apart from one chorus which is in the Nationalbibliothek, Vienna), but was later rediscovered and is now in the Paris Opera Library. Extracts from it have been published by Jean Philippe Chantavoine in " Die Musik," Vol. III, No. 16 (1904), and reproductions of the costumes by M. D. Calvocoressi in his " Franz Liszt " (Paris, 1905). The music shows a real melodic gift and a considerable power of characterisation; but the opera was written and put on more for propaganda purposes than anything else, and one cannot imagine that Liszt felt much interest in a conventional plot of this kind. It was not a great success, and there were only four performances.

Liszt completed his next original work in 1834; and the six years after 1828 marked the transition to maturity. External events wrought a great change in his character; first his illness in 1828-9, and the outbreak of religious fervour which accompanied it, gave him a strong distaste for the career of a travelling virtuoso for which he seemed to be destined; then the revolution of 1830 infected him with romantic revolutionary ardour; at the same time his entry into contact with the literary and artistic world of Paris at the moment of the birth of the Romantic Movement revealed to him how defective his general education had been—a defect which he determined to remedy as speedily as possible; and finally the successive impacts of Berlioz, Paganini and Chopin gave him a completely new musical outlook. This rapid series of events, coming at a time when adolescence was giving way to maturity, completely altered the whole course of his life and work. Hitherto he had been merely a successful infant prodigy, composing in his spare time works bounded by his technical powers and partly (at any rate) influenced by the views of his teachers; from now on his horizon was incomparably widened in all directions. Berlioz was of course to affect him more considerably later on, in his orchestral works; but already he derived from him the force and vehemence of thought, the nervous and diabolical energy, the feeling of wild and incoherent ambition, the striving after " Babylonian grandeur," which can be found only a few years later in the earliest versions of the *Vallée d'Obermann* and the *Dante Sonata*, as well as in the *Grandes Études* of 1838.

From Paganini he derived, first of all, a new transcendental technique which was as far in advance of anything previously written for the piano as Paganini's was in advance of anything written for the violin; secondly, as Mr. Sacheverell Sitwell suggests, that sense of showmanship (in the best meaning of the word) which enabled him to overcome his worst enemies once he sat down at the piano; and thirdly that feeling of Mephistopheleanism and diabolism which both Paganini and Berlioz had in common. The personality of Paganini was held in almost superstitious terror during his lifetime; and though his compositions sound harmless enough to us to-day, there is something in their chill perfection which makes one realise that they could well have excited a feeling of awe among contemporary hearers. A whole section of Liszt's works expresses this kind of feeling; it includes the Faust Symphony, the Totentanz, the Dante Sonata, the Mephisto Waltzes and a great number of the late piano pieces. Indeed Liszt's whole personality in later years took on a diabolical flavour; one remembers Gregorovius's description of him in 1865 as "Mephistopheles disguised as an Abbé."

But the composer who was perhaps to have the most immediate influence on him in these years was Chopin. They were of the same age and in similar circumstances; they came from similar environments, for in both Poland and Hungary there was no bourgeois class, and any talented artist tended to be drawn into the aristocratic salons in order to make his career. Chopin had had the advantage of a better upbringing than Liszt, and was more experienced; what he chiefly showed to Liszt was the poetical approach to music. Liszt had been reading intensively in the last few years, and the effect of Chopin was to make him connect music more closely with the rest of his thought. Instead of writing music to fill the academic forms he had been taught in his youth, he now began to see every piece as the musical expression of a certain idea or state of mind, sometimes derived from literature or art, sometimes from experience. It is this, chiefly, that has caused him to be misjudged by those who regard all music as "abstract" and therefore incapable of portraying a mood or an idea; to Liszt, Chopin and the Romantic Movement in general, on the other hand, it was part of the essence of music to be able to do this. Debussy rightly wrote in *Monsieur Croche Antidilettante*: "The undeniable beauty of Liszt's work arises, I believe, from the fact that his love for music excluded every other kind of emotion. If sometimes he gets on easy terms with it and frankly takes it on his knee, this is surely no worse than the stilted manner of those who behave as if they were being introduced to it for the first time; very polite but rather dull. Liszt's genius is often disordered and feverish, but that is better than rigid perfection, even in white gloves." In the years 1829-34, then, Liszt was gradually developing a totally new approach to music. He began with two brilliant Fantasies, one on themes from Auber's *La Fiancée*, written in 1829, which already shows complete command of virtuoso technique and much originality, and one on " La Clochette " (La Campanella) of Paganini, written in 1832 under the immediate impact of hearing Paganini play—to this we must return when considering the Paganini studies in general. Between the two came the idea for a far more interesting work, a Revolutionary Symphony, inspired by the July Revolution of 1830. According to Liszt's " official " biographer Lina Ramann, Liszt took as his example the *Battle of Vittoria* of Beethoven, and like his model, intended to introduce national themes into his work. These were to include a Hussite chorale from the fourteenth century, *Ein Feste Burg* and the Marseillaise; this choice of revolutionary songs from the Slavonic, Teutonic and Latin worlds no doubt symbolised the universal brotherhood of revolution against oppression. The work was never completed, though some of it was later used in the symphonic poem *Héroïde Funèbre*; however some sketches of it are preserved in the Weimar Liszt Museum, and some of these are reproduced in facsimile in the first volume of Peter Raabe's "Franz Liszt" (Stuttgart, 1931). At the side of the music Liszt sketched out a " programme," which, in so far as it can be deciphered, reads as follows: " indignation, vengeance, terreur, liberté! désordre, cris confus (vague, bizarrerie), fureur . . . refus, marche de la garde royale, doute, incertitude, parties croisantes . . . 8 parties différentes, attaque, bataille . . . marche de la garde nationale, enthousiasme, enthousiasme, enthousiasme! . . . fragment de Vive Henri IV dispersé. Combiner ' Allons enfants de la patrie '." The sketch is headed " Symphonie " and is dated " 27, 28, 29 Juillet—Paris." Musically the sketches reveal very little, though a clear reference to the Marseillaise is apparent; but it is interesting to note that twenty years later, presumably as the result of the European uprisings of 1848-9, Liszt felt impelled to take up the project again, this time in the form of a five-movement symphony. The first was to be *Héroïde Funèbre*, the second a setting of "Tristis est anima mea,"* the third was to be based on the Rákóczy and Dombrowski marches (symbolising Hungary and Poland), the fourth on the Marseillaise, and the last movement was to be a setting of Psalm 2, "Quare fremuerunt gentes?" From this whole plan only *Héroïde Funèbre* was completed; but fragments of both the earlier and later versions appeared separately; the Hussite chorale, *Vive Henri IV*, and the Marseillaise were arranged for piano, and we find *Ein Feste Burg* both in the "Huguenots" Fantasy and in Liszt's transcription of Nicolai's *Kirchliche Festouverture*; in addition the Rákóczy March was arranged both for piano and for orchestra.

Before attempting any more original works, Liszt now turned his attention to transcriptions and fantasias. He first set himself what might seem the completely impossible task of transcribing Berlioz' Symphonie Fantastique for piano—surely one of the most unpianistic works ever written! Yet Sir Charles Hallé in his memoirs relates an occasion, at which he was present, when Liszt played his transcription of the March to the Scaffold immediately after an orchestral performance of the

* Sketches for both this and Psalm 2 exist at Weimar; "Tristis" is partly in the Hungarian style, and has no musical connection with the setting of the same words in "Christus."

same piece, and received even more applause than the orchestra. Looking at Liszt's "partition de piano," as he called this type of transcription, one can well believe it, for it is astonishingly well done; Liszt has not simply arranged the notes for piano, but has recast the texture in such a way as to make the piano give an orchestral effect. The extract from the March to the Scaffold on p. 7 indicates the type of writing he used. The purpose of the transcription was of course to help Berlioz at a time when he found it difficult to get orchestral performances of his works: Liszt not only played it in his own concerts, but actually bore the expenses of its publication, so that it could reach as wide a public as possible. He followed this up by writing a short piano piece, *L'Idée Fixe, Andante amoroso*, on the main theme of the Symphony, and also by transcribing Berlioz' overture *Les Francs-Juges*. Next year (1834) he wrote a Grande Fantaisie Symphonique on themes from *Lélio* for piano and orchestra, and in 1836 he also transcribed *Harold in Italy* and the *King Lear* overture.

Liszt has often been attacked for writing these transcriptions and fantasies, and it is as well to discuss this question before we proceed further. To begin with, these works fall into two main classes: the "partitions de piano," which are more or less straight transcriptions from one medium to another, and the fantasies, which are original works based on other composers' themes. There are of course borderline cases, where a transcription is so embellished with added detail that it almost becomes an original composition; but in general these two classes hold good. The purpose of the "partition de piano" was normally to help the composer by making his orchestral works more easily accessible to a wider audience. Berlioz, as we have seen, was the first whom Liszt helped in this way; but many others followed including Wagner, Glinka, Gounod, Saint-Saëns, Cui, Dargomijsky and a number of the younger German composers. In addition Liszt was able to make the works of Bach, Beethoven and Schubert known at a time when they were insufficiently appreciated by the concert audiences of the day. These transcriptions therefore served approximately the purpose of the modern gramophone record, that of presenting in a convenient form works which would not otherwise be easily available; and there is no doubt that Liszt's transcriptions usually give as good an idea of a work as is possible in a totally different medium. It is true that some of his transcriptions of Beethoven and Schubert contain a good deal of added decoration which may not always be felt to be in keeping with the original; but in most cases Liszt also gave a simpler form as an *ossia* for those who prefer it, and his transcription of Schubert's Wanderer Fantasia for piano and orchestra, for instance, is generally agreed to be a brilliant realisation of an extremely difficult task. It is also true that many of the works transcribed by Liszt, at all periods of his life, are by third- or fourth-rate composers; many of these were transcribed either as compliments to Liszt's aristocratic friends, or for the benefit of young and comparatively unknown composers to whom the name of Liszt on the cover of their pieces would be a great help. A further point is that in Liszt's later years payments from his publishers represented practically his entire source of income; he had given up playing the piano professionally, and he never accepted a fee for teaching. He may have wasted a good deal of time on some of these transcriptions, but at least they provided an outlet for his abounding energy.

The fantasies are a different matter. Many, but by no means all, of these are on operatic themes—for instance, among the earliest are those on *La Campanella*, *Lélio* and Mendelssohn's Songs Without Words—and it is true that brilliant fantasies on current operatic successes were part of the stock-in-trade of every fashionable pianist of the day. We may look down our noses at them now, but we must remember that taste was very different in the 1830s, and who can blame Liszt for wanting to play the game even better than his rivals? Even César Franck in his younger days wrote a number of fantasies, including one on " God Save the Queen " for piano duet, as well as a " Souvenir d'Aix-la-Chapelle "; it was in fact the rule rather than the exception to do so. The sole criterion is the musical value of the result; and here I think we must admit that Liszt not only outstripped every other competitor in the field, but produced a number of real winners. The best of the fantasies are formally well organised, musically beautiful and exciting, and of course brilliantly written—one has only to compare them with those of, say, Thalberg to see the difference between a first-rate and a second-rate mind. Naturally there are bad fantasies as well as good—here again Liszt wasted a good deal of time on inessentials—but it is ridiculous to condemn all the fantasies out of hand as mere drawing-room fireworks. This will become apparent when we come to discuss them in detail.

The Grande Fantaisie Symphonique on themes from Berlioz' *Lélio* was written, as we have seen, in 1834, and was first performed in Paris in the following year. The MS., which is unpublished, is in the Liszt Museum at Weimar. It seems unlikely that the scoring is by Liszt himself; at that time he had little knowledge of orchestration, and the first work which he is known to have scored himself is the first Beethoven Cantata of 1845. This is borne out by the fact that the score is written in another hand, but contains additions and alterations in Liszt's writing, and even some pencilled comments, such as " Bon!," which appear to express approval of some idea of the orchestrator's which Liszt had not thought of himself. The Fantasy is in two sections, the first based on the setting of Goethe's *Der Fischer* (*Le Pêcheur*), and the second mainly on the *Chanson de Brigands*. The opening section, *Lento*, is a poetical meditation on Berlioz' theme, with alternating moods of animation, despair, calm, delicacy—in fact the real essence of *Lélio* as a whole seen in miniature; the second section, in which the trombones enter for the first time, is naturally more dramatic and exciting, though it is interrupted just before the end by a return of the *Lento* theme. The Fantasy is in fact a re-creation of Berlioz' work in Liszt's terms—an unusual tribute from one composer to another, perhaps, but certainly a practical one in the case of a work which needs such elaborate and unusual resources in its original form. The Fantasy is one of the most clearly written and successful of Liszt's early works, and obviously deserves publication and performance.

The *Grosses Konzertstuck* on themes from Mendelssohn's Songs without Words for two pianos dates from exactly the same time as the *Lélio* Fantasy, and was in fact first performed at the same concert. Liszt had met Mendelssohn a few years earlier and no doubt wished to do a similar service to what he had done for Berlioz. The work has remained unpublished;

it is Liszt's first and only original composition for this medium, the later *Concerto Pathétique* being a revised version of the *Grosses Konzertsolo* for solo piano. The same year, 1834, saw Liszt's real beginning as an original composer, and from it date four extremely remarkable works. The first of these is the single piece *Harmonies Poétiques et Religieuses*; the title is derived from a collection of poems by Lamartine (1790-1869), and in front of his piece Liszt quoted Lamartine's preface to the collection, which is worth reproducing in part, for it shows Liszt's new approach to his art. It begins: " These lines are only addressed to a small number," and continues later: " There are hearts broken with grief, rebuffed by the world, who seek refuge in the world of their thoughts, in the loneliness of their souls, to weep, to wait or to worship; may they be visited by a Muse solitary like themselves, find sympathy in her harmonies, and sometimes say while listening to her: we pray with thy words, we weep with thy tears, we invoke with thy songs! " The piece is a kind of free improvisation, mostly without time or key signature, and far bolder than any previous attempts in the same direction, such as Beethoven's Fantasia, Op. 77, or the slow movement of Mendelssohn's early piano sonata in E minor (1826). In this piece we can see Liszt trying to reproduce the effect of his own playing in a more minute way than had ever been attempted before. The beginning is marked " avec un profond sentiment d'ennui," and the mood ranges through a recitative passage and an *Agitato assai* to an *Adagio* and an *Andante religioso*, ending gloomily, yet inconclusively, with a bass recitative. The final section, based on this theme, which in various forms runs throughout the work—an anticipation of Liszt's later

method of " transformation of themes "—has a strong feeling of the late Beethoven, an influence which only appears occasionally in Liszt's work, particularly in the *Benediction of*

God in Solitude in the later set of pieces called *Harmonies Poétiques et Religieuses* (1845-52). This section also contains some remarkable chromatic harmony, such as the following:

But it is unfair to judge this piece by extracts alone; it is well worth studying in its entirety, for it gives a very clear insight into Liszt's methods of thought. It is in any case an extremely remarkable work to have been written within seven years of the death of Beethoven, especially by a composer of only twenty-three.

The piece was originally conceived for piano and orchestra; in the later set of *Harmonies Poétiques et Religieuses* Liszt included it in a revised version under the title *Pensée des Morts*, and disclaimed the earlier version as " tronquée et fautive." Few will agree with him in this, however, for the later version is considerably inferior to the first, and contains a good deal of that rather stifling atmosphere which has somehow turned sour most of the set of *Harmonies*. The revision is interesting, however, as showing the connection between the original version and another piece written at the same time, the " instrumental psalm " *De Profundis* for piano and orchestra, of which the unfinished MS. is in the Liszt Museum. It is dedicated to the Abbé Félicité de Lamennais, the philosopher and religious and social thinker, who had a considerable influence on Liszt at this time; Liszt was in fact staying with Lamennais at La Chesnaie in Brittany during the summer of 1834. *De Profundis* begins with a long, wild and stormy introduction for piano and orchestra; the piano part is brilliant, but the orchestration, which is in Liszt's own hand, is only sketched in in the most bare and simple manner. After a series of repeated held chords the piano announces the main psalm theme, the words of Psalm 130 being written against it in the score; it was this theme and these words which Liszt later inserted into *Pensée des Morts*. A good deal of the work consists of a kind of struggle between the psalm theme and the stormy motives of the introduction; later there is a fairly long section in a lighter and more lively mood; then after a partial repetition of the opening themes the work ends with a section in the style of a march, containing a new version of the psalm theme. The work is a sketch in the sense that the orchestral part is far from complete; but there is something written in every bar, and with a little ingenuity it might be possible to complete the score. It is certainly a most remarkable and interesting product of Liszt's youthful romanticism, wild and chaotic though a good deal of it is.

Another work connected with and dedicated to Lamennais and dating from this period is *Lyon*, a stirring march-like piece which was inspired by a workers' uprising in that city; the dedication implies that both social and religious problems were discussed by Liszt and Lamennais. *Lyon* was published as No. 1 of the *Album d'un Voyageur*, the earliest form of the *Années de Pèlerinage*, and further description of it must be deferred until these collections are discussed as a whole.

The final work of this group is the set of three *Apparitions* for piano. These also date from 1834; the title was suggested * by a set of visionary songs, called *Auditions*, by Christian Urhan, a violinist of German origin living in Paris with whom Liszt often made music—in 1837, together with the 'cellist Batta, they gave four evenings of Beethoven's chamber music, not a very popular proceeding in the Paris of that time! These poetical pieces strike a slightly more secular note than the others so far discussed, in that the melodic lines are more similar to the *bel canto* of the Italian opera—an influence which was to remain with Liszt throughout his life. Nevertheless the pieces show extraordinary delicacy as well as strong romantic feeling; the first in particular is a little masterpiece, and Busoni rightly characterised it as " romantic, emotional, philosophical, and possessing that breath of Nature which in art is achieved so rarely and with such difficulty." The second he described as a " capricious, almost conversational piece, and an example of that ' subjective impressionism ' which Schumann aimed at and achieved in his earlier works; but Liszt is a cosmopolitan,

---

* An alternative suggestion (perhaps more plausibly) relates these pieces to Lamartine, one of whose poems is called *Apparition*.

speaking to all men of taste, where Schumann remains a German." The third is a somewhat complex fantasia on a waltz of Schubert, and was later transformed into the fourth of the *Soirées de Vienne*; its character is sufficiently indicated by its markings—" molto agitato ed appassionato, vibrante, delirando, precipitato, avec coquetterie, religiosamente, con gioja," etc. It is interesting as being Liszt's first work on a theme by Schubert, who had been dead only six years, and was more or less completely unknown outside Vienna. The three *Apparitions* have all the glow of the early Romantic Movement; one cannot say that they show any direct influence either of Chopin or of Schubert, but they do constitute a tribute to the spirit of these composers.

During the rest of this period (1835-9) Liszt occupied himself with three main types of composition: the two sets of Études (Transcendental and Paganini), the earliest versions of the first two books of the *Années de Pèlerinage*, and a number of lighter pieces, transcriptions and fantasies. The earliest version of the Transcendental Studies dates back to Liszt's sixteenth year; it was called, as we have seen, *Étude en 48 Exercices*—though in fact only twelve were ever written. Liszt, presumably following the example of Bach, intended to write two studies in each major and minor key, and the twelve completed studies are arranged in a definite key sequence —C major, A minor; F major, D minor; B flat major, G minor, etc. This first version was dedicated to Mlle Lydie Garella, a young lady in Marseilles with whom Liszt used to play piano duets. It cannot be said that these studies are particularly interesting in themselves; they are written more or less in the style of Liszt's master Czerny, and their chief value resides in what happened to them later. For the *12 Grandes Études*, published in 1839 (here twenty-four were announced, but again only twelve completed), present these same simple pieces in a fantastically transformed and enlarged form. They are the result of Liszt's new transcendental technique, and they bristle with the most hair-raising technical difficulties. Before discussing the studies in detail, it is best to complete the tale of their later history. In neither the 1826 nor the 1839 versions do the individual pieces bear any titles; but in 1847 Liszt published the fourth study separately with the title *Mazeppa* and a slightly altered ending to fit the well-known story treated by Byron and Victor Hugo, among others. Finally, in 1852 Liszt republished the whole set in a revised form, under the title *Études d'Exécution Transcendante*, and this is the form in which they are best known to-day. In this version titles are added to all but two pieces; the list is as follows:

1. Preludio.	7. Eroica.
2. in A minor.	8. Wilde Jagd.
3. Paysage.	9. Ricordanza.
4. Mazeppa.	10. in F minor.
5. Feux Follets.	11. Harmonies du Soir.
6. Vision.	12. Chasse-neige.

It should be added that both the second and third versions bear a dedication to Czerny.

Comparison of the three versions of these studies makes a fascinating study, and gives an admirable insight into Liszt's methods. (All three versions are published in the Breitkopf Collected Edition.) The differences between the last two versions are not great; apart from the addition of the titles, they consist mainly in a certain amount of cutting, and smoothing out of abnormal technical difficulties, which abound in the second version. Musically, in fact, the 1839 version presents the studies in all their essentials; as a simple example of the kind of relation that exists between the three versions we may quote the beginning of No. 8 (Wilde Jagd) (Ex. 4, p. 16).

Berlioz wrote of Liszt's music about this time: " Unfortunately one cannot hope to hear music of this kind often; Liszt created it for himself, and no one else in the world could flatter himself that he could approach being able to perform it." Compare this with Professor Dent's description of Busoni playing Liszt: " The greater works of Liszt, which minor pianists turn into mere displays of virtuosity because their technique is inadequate for anything beyond that, often sounded strangely easy and simple when they were played by Busoni. The glittering scales and arpeggios became what Liszt intended them to be—a dimly suggested background, while the themes in massive chords or singing melodies stood out clear." Liszt evidently realised the danger of the extreme technical difficulty of some of his works, and therefore revised

them in later life; but even so a great deal of his music is in the
unfortunate position of being playable in the way Liszt meant
it to be played by only a handful of pianists in each generation,
while remaining at the mercy of every pianist who has enough

technique to play the notes and nothing more; and thereby
the misleading impression, to which we are so well accustomed,
is perpetuated. Liszt did not invent his transcendental
technique merely in order to dazzle his hearers and show that
he was a better pianist than his rivals; he did it because he was
thereby able to draw new and almost orchestral effects from the
piano, which incomparably widened its range of expression—
and all subsequent composers for the piano are grateful to him.

The individual studies may be briefly discussed. The first,
Preludio, is a mere " warming-up," a running of the pianist's
hands over the keys so that he may get used to his instrument—
a thought typical of a practical pianist-composer, but at the
same time musically very effective. Then the real business

begins with the A minor study, a violent and brilliant piece, inspired perhaps by Paganini. *Paysage* is a calm and beautiful landscape scene, simply, yet originally expressed; some of the syncopated progressions in the middle have more than a foretaste of Brahms—who was five when this piece was written!

Ex.5

Mazeppa has of course become a pianists' war-horse; it is certainly brilliant and effective, but the thought is hardly subtle. *Feux Follets*, also well known, can never become hackneyed because of the ingenuity and delicacy with which it is carried out. *Vision* is a broad and impressive piece, showing how Liszt could admirably fill an ample canvas when he wished. In *Eroica* (of which the Introduction is derived from the early *Impromptu on themes of Rossini and Spontini*) we have a fine example of that " vibrating of the heroic string " which is found in many of Liszt's works, particularly those in the Hungarian manner. In *Wilde Jagd* we have all the feeling of the romantic nocturnal hunt, celebrated in all German music from Weber onwards; *Ricordanza* is beautiful, nostalgic, " like a packet of yellowed love letters," as Busoni described it. Its main theme raises an interesting problem; as it stands, one would immediately call it Chopinesque, but when one realises that in the 1826 version it actually appears in this form:

Ex.6

it is clear that this cannot be the explanation. At this time Chopin was only sixteen, and Liszt could not have known his music; Field, Hummel or Dussek are possible influences, but it seems more probable that both Liszt and Chopin derived this type of *cantilena* from the Italian opera, or perhaps that it was natural to them both independently from their earliest years.

The three remaining studies need little comment. No. 10 is again a violent, rather savage piece; *Harmonies du Soir* admirably conjures up an atmosphere of evening calm, while *Chasse-neige*, in some ways the most impressive of the whole set, gives the gloomy, rather sinister impression of a whole landscape being slowly and relentlessly covered with snowdrifts. Liszt's Transcendental Studies represent his first major mature collection of pieces, his challenge to the world, as it were; not all the pieces are masterworks by any means, but the collection as a whole is an impressive one. In it Liszt not only showed himself as the creator of a new and revolutionary technique, but also through his expressive powers clearly marked himself as one of the leaders of the Romantic School.

In 1832, as we have seen, Liszt wrote the *Grande Fantaisie de Bravoure sur la Clochette de Paganini* (on the theme of the rondo from Paganini's Concerto No. 2 in B minor, Op. 7); this was followed in 1838 by the six *Études d'Exécution Transcendante d'après Paganini*, of which the third is again *La Campanella*, but on a far smaller scale; the remaining five are transcriptions of some of the caprices for solo violin, which Paganini had been persuaded to publish about 1830. Twelve studies were originally announced for publication, and possibly Liszt may have intended at one time to transcribe all the twenty-four—according to Lina Ramann he began this task at about the same time as the *Clochette* fantasy—but no trace of the others has been found. In 1851 a revised and simplified version of the six studies was published—the form in which they are generally known to-day—as the " seule édition authentique, entiére-ment revue et corrigée par l'auteur." Later the three different forms of the fourth, so-called " Arpeggio " study—it had been printed in two separate versions in the 1838 edition—were published with a preface by Liszt's pupil Eduard Reuss; and between 1913 and 1923 Busoni published his own versions of the whole set, so that these études now bear the names of the three greatest virtuosi in history.

This is not the place to discuss Paganini's position in musical history as a composer—readers may be referred, for instance, to Bernard van Dieren's brilliant and characteristic appreciation of him in " Down Among the Dead Men " (pp. 268 *seq.*)— but it is clear that Liszt did not undertake these transcriptions merely because of the technical problems involved—he certainly had a very real interest in Paganini's music for its own sake, and in fact felt a kind of spiritual kinship with him. Nor was he the only composer to be attracted by Paganini; Schumann, Brahms and, in our own day, Busoni, Rachmaninoff, Tommasini, Blacher, and Lutosławski—to name only a few—have also felt a similar attraction. The variations of Brahms, Rachmaninoff and Tommasini (*La Carnevale di Venezia*) are not merely " on a theme of Paganini," but owe their virtuosic style entirely to the influence of his works. Busoni transcribed parts of two of the caprices to form the fourth book of *An die Jugend*, and included two of his Paganini-Liszt transcriptions in his *Klavierübung*. Schumann transcribed twelve of the caprices in two sets, published in 1833 (Op. 3) and 1835 (Op. 10), respectively. It is interesting to see how completely they are Schumannised—for while Schumann's version is far nearer to the actual notes of the original than Liszt's is, the spirit is entirely different. Schumann in his introduction to the first set says: " Although the interest which the composition itself had for the Editor incited him to the work (of transcription), he also designed, by means of it, to give solo players an opportunity of removing a reproach often cast at them, that they make too little use of other instruments and their peculiarities for the improvement and enrichment of their own; but chiefly he hoped to become useful to many estimable artists, who from fear of all that is new are unwilling to forsake antiquated rules." His approach was thus not so very different from Liszt's, who, while dedicating both versions of his Paganini studies to Clara Schumann, paid her husband the compliment (Busoni, perhaps inaccurately, calls it a " weltmännischer oder mephistophelischer Laune ") of printing his own version of the G minor " Tremolo " étude side by side with his own new and far more brilliant transcription, as if to show superiority over even Schumann's new advances.

The *Clochette* Fantasy of 1832 (incidentally the Liszt Museum possesses the unfinished sketch of another work on the same theme), like many of Liszt's works of this period, appears, from the direction " Tutti " which occurs occasionally in the first printed edition of Mechetti, to have also existed in a version with orchestral accompaniment. It is a terrific piece of virtuosity, and it says much for Liszt's powers of assimilation that he could invent and perfect a new technique of such unheard-of brilliance and difficulty within such a very short time of hearing Paganini play. It consists of a slow, free introduction, a rather capricious and violent bridge passage working up to a fairly straightforward statement of the theme; then follows a " Variation à la Paganini " which reproduces some of Paganini's effects in harmonics; the work ends with a " Finale di Bravura," beginning with the theme in the major and again working up to a climax. As Busoni said, " in spite of much monstrosity (Ungeheuerlichkeit) the work shows an original mind peeping through, a constrained emotion struggling for expression." Liszt in later years noted that the working up in the final variation was an anticipation of several passages in the symphonic poems; and if the work as a whole exploits Liszt's new-found technique too relentlessly to be fully tolerable either for present-day pianists or audiences, it is still worth an occasional revival for the sake of the many beautiful and poetical passages which it contains.

The Paganini études proper are short pieces, on a smaller scale than the Transcendental Studies, but extremely concise, and (in the 1838 version at any rate) bristling with difficulties. They are in fact, in this version, too difficult to be playable comfortably even by first-class pianists, and in 1851 Liszt revised and simplified them drastically. (Raabe (II, 29) gives an interesting example of his rearrangement of a chord of B flat in the two versions.) Schumann, not unnaturally, was inclined to be suspicious of the earlier version, and remarked that the musical content often had no relation to the mechanical difficulties involved, and that Liszt himself would certainly have to practise these pieces before he could play them; but, as Busoni said, the transcriptions are of real **Paganinesque** *diablerie*—these études possess the trenchant, yet expressive, realism of Paganini's own playing.

To his first étude, the " Tremolo " study, Liszt prefaced the quasi-improvisatory cadenza, another " running of the hands over the instrument," as it were, which really belongs to Paganini's fifth caprice in A minor—this is Schumann's Op. 3, No. 1. It is interesting to make a comparison of the different versions of this study; on page 22 is shown the return to the main theme.

It will be seen that Liszt's 1851 version, though considerably simplified, is just as brilliant as the 1838 version—the difference is that between experimentation and mature mastery. As Raabe notes (II, 30), Liszt's penchant for crossings of the hands in the first version often produced striking effects, though not always; and such of these experiments as were unsuccessful were altered in the final version. But it is still possible to prefer the first version in some ways, as Mr. Sitwell does, for instance, and it is significant that Busoni in his edition has frequently restored what Liszt pruned away in 1851. Thus the fourth étude was written by Paganini in arpeggios across the strings; Liszt issued it in 1838 in two forms, the first mostly in two- and three-part chords divided between the hands, the second often with four-part chords in one hand and a separate counterpoint, together with its accompaniment in chords, in the other; in 1851 Liszt wrote it on one stave only, chiefly in single notes divided between the hands, and Busoni again enlarged it by writing it in octaves divided between the hands.

The second étude is the so-called " Octave " study, Paganini having written its middle section as a study in rapid octave playing; the third is " La Campanella," transposed down to G sharp minor from the A minor of the *Clochette* fantasy; a pianist would find it easier to make frequent leaps up to a high D sharp rather than an E, whereas Paganini imitated the sound of a little bell by harmonics on his open string. The fifth study is " La Chasse," a very successful imitation of fifes and hunting horns; here again the 1851 version is nearer to Paganini than that of 1838, which is unnecessarily complicated, and misses the effect of the hunt dying away in the distance by adding a frenetic coda. The last étude is the " Theme and

"Variations," on which Brahms, Rachmaninoff and others have based their works; it is the largest in scale of the set, and remains a fine piece of virtuosity even in the later version.

There could be no greater contrast to the two brilliant sets of studies that we have been discussing than the poetical pieces which make up the *Album d'un Voyageur* and the first two books of the *Années de Pèlerinage*. As the history of these collections is somewhat complex, it will be as well to set it out in detail before proceeding further. The pieces included in the *Album d'un Voyageur* were mostly written in 1835-6, that is to say during Liszt's travels with the Comtesse d'Agoult in Switzerland. The whole collection was published in 1842, and contains the following pieces;

I. Impressions et Poésies.
    1. Lyon.
    2a. Le lac de Wallenstadt.
    2b. Au bord d'une source.
    3. Les cloches de G. . . .
    4. Vallée d'Obermann.
    5. La chapelle de Guillaume Tell.
    6. Psaume.
II. Fleurs mélodiques des Alpes (without titles).
    Nos. 7a, 7b, 7c, 8a, 8b, 8c, 9a, 9b, 9c.
III. Paraphrases.
    10. Ranz de vaches (Aufzug auf die Alp, Improvisata).
    11. Un soir dans les montagnes (Nocturne pastoral).
    12. Ranz de chèvres.

The confusion, such as there is, arises from the fact that the various sections of this collection had previously been published separately under different titles; the *Impressions et Poésies* first appeared about 1840 as " Ire Année de Pèlerinage, Suisse," and the *Fleurs mélodiques* as " 2de Année " (the " 3me Année " at this stage consisted of the first seven of the *Magyar Dallok*, which will be discussed later in connection with the Hungarian Rhapsodies). In addition No. 7c appeared separately as " La Fête villageoise." The *Paraphrases* had been published as early as 1836 under the title " 3 Airs suisses "; Liszt did not include these in the later volumes of the *Années de Pèlerinage*, and they were republished in 1877 as " 3 Morceaux suisses." The final works in this group are the *Fantaisie romantique sur deux mélodies suisses* (which is based on the same themes

as No. 7b) and the *Eglogue*; both these were written and published in 1835-6. (For further details see Raabe II, 244-5,
and the Liszt catalogue in " Grove's Dictionary of Music.")

Between 1848 and 1853 Liszt revised the whole collection,
and issued it in 1855, in the form in which we know it to-day,
as *Années de Pèlerinage, Première Année, Suisse*. The contents of
this volume are as follows:

1. Chapelle de Guillaume Tell.
2. Au lac de Wallenstadt.
3. Pastorale.
4. Au bord d'une source.
5. Orage.
6. Vallée d'Obermann.
7. Eglogue.
8. Le mal du pays.
9. Les cloches de Genève.

The only completely new piece in this collection is the *Orage*;
for the *Pastorale* and *Le mal du pays* are new versions of Nos.
7c and 7b respectively of the *Fleurs mélodiques*; and *Un soir
dans les montagnes* from the *Paraphrases* also contains a description
of a mountain storm, of a rather similar character to *Orage*.

As we have seen, the *Paraphrases* were republished in 1877;
and therefore the only pieces from the *Album d'un Voyageur*
which Liszt finally jettisoned were *Lyon*, the *Psaume*, and some
of the *Fleurs mélodiques*. (All these pieces have been republished
in the Breitkopf Collected Edition.) We must certainly regret
the loss of *Lyon* from the later collection, as it is a very fine and
characteristic piece. It is prefaced with the sentence " Vivre
en travaillant ou mourir en combattant "—Life in work or
death in fighting—and, as we have seen (p. 13), was connected
with Liszt's support of the revolutionary movements of the time.
It rises inevitably to a fine climax, and is certainly worth the
attention of pianists. The remainder of the *Impressions et
Poésies* will be discussed in conjunction with their later versions,
apart from the *Psaume de l'Église a Genève*, a straightforward
and not particularly remarkable transcription of a setting of
Psalm 42—" As the hart panteth after the water-brooks."

The *Fleurs mélodiques* are a set of fairly short pieces based
mainly on Swiss folk tunes (further details about Liszt's use

of these tunes will be found in the preface to the Breitkopf Collected Edition, Piano Works, Vol. IV); though there are moments of elaboration, on the whole they are simple and straightforward, and the two which were taken over into the *Années de Pèlerinage* are certainly the most interesting. Of the *Paraphrases*, the *Ranz de vaches* and *Ranz de chèvres* are based on themes by the Swiss composer F. F. Huber (1791-1863); the former is in the form of variations, with some interludes, while the latter is a rondo. Both are cheerful, pleasant pieces in which a pastoral character is combined with brilliant writing. *Un soir dans les montagnes* is rather more elaborate; it begins and ends with a quiet, pastoral section based on a theme by the Basle publisher E. Knop, while the central portion is a vivid storm scene—this piece too is well worth revival for its picturesque atmosphere.

The *Fantaisie romantique sur deux mélodies suisses* is a very much more extended and elaborate piece than the foregoing. As we have seen, it is partly based on the same themes as *Le mal du pays*, but the rather curious and restrained character of the latter piece is entirely absent from the Fantasy, which is a brilliant, somewhat overwritten work and not one of Liszt's greatest creations. Nevertheless it does catch something of the same fresh atmosphere as the rest of the Swiss pieces, which are unique in Liszt's output in that they make a consistent and on the whole successful use of the folk-material of a country with which Liszt had no direct personal link (apart from the circumstances of his travels there); yet one feels that in these works Liszt had thoroughly absorbed the atmosphere of his new surroundings, and they never fall into mechanical pastiche of the " picturesque " type.

We can see this approach at its best in the final version of the *Années de Pèlerinage*. In nearly every case Liszt has tightened up the earlier versions and made the pieces more consistent and satisfactory, without losing the freshness of the original inspiration. *The Chapel of William Tell*, with the motto " One for all, all for one," is a fine, straightforward portrait of the Swiss national hero, in which we hear the trumpets of revolution echoing through the mountain valleys. (Why is this piece so rarely played?) *Au lac de Wallenstadt* bears a quotation from Byron's " Childe Harold ":

> . . . thy contrasted lake,
> With the wild world I dwell in, is a thing
> Which warns me, with its stillness, to forsake
> Earth's troubled waters for a purer spring.

Comtesse d'Agoult wrote in her memoirs: " the shores of the lake of Wallenstadt kept us for a long time. Franz wrote there for me a melancholy harmony, imitative of the sigh of the waves and the cadence of oars, which I have never been able to hear without weeping." It is certainly one of the most beautiful and successful pieces in the whole collection, and it is significant that Liszt preserved it without change in the later version. The next piece, *Pastorale*, is also of a charming simplicity; it is a somewhat shortened version of *La Fête Villageoise* from the *Album d'un Voyageur*, and Liszt has definitely improved it by omitting a central section in Ländler tempo. As it stands it has all the freshness and spontaneity of Bartók's folk song arrangements; and, as we shall see, in some of his versions of Hungarian folk songs Liszt actually anticipated Bartók's work in this field.

The next piece, *Au bord d'une source*, is the only one of this collection which has got into the general repertoire. Prefaced by a quotation from Schiller, " In murmuring coolness the play of young Nature begins," it preserves all the limpid freshness of its subject, and its popularity is well deserved. It is interesting to note the differences in the layout for the hands between the earlier and later versions; in 1835 Liszt wrote:

In the later version, by crossing the hands, a far more subtle effect is obtained:

Orage, as its title implies, is a graphic description of a mountain storm scene. The opening, as we shall see, has some parallels with the so-called *Malediction* concerto. For the rest, it is

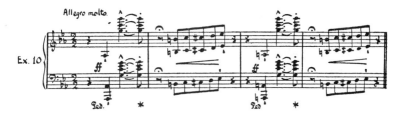

Ex. 10

typical of Liszt in his more violent mood; the main theme is of a type which we come across frequently in the symphonic poems and other works of the Weimar period.

Ex. 11

Here again is a piece which could well be revived by pianists.

The sixth piece, *Vallée d'Obermann*, is the most considerable in the collection; it is prefaced by a long quotation from the novel *Obermann* by Etienne Pivert de Senancour (1770-1846), to whom Liszt dedicated his piece. This romantic work had a considerable influence on Liszt at the time; this quotation will give some idea of its character. " Vast consciousness of a Nature everywhere overwhelming and impenetrable, universal passion, indifference, advanced wisdom, voluptuous abandon, all the desires and all the profound torments that a human heart can hold, I have felt them all, suffered them all in this memorable night. I have made a sinister step towards the age of enfeeblement; I have eaten up ten years of my life." The work has a most curious atmosphere, gloomy, sinister and resigned by turns, but ending with a real pæan of joy; it is in some ways a kind of earlier *Verklärte Nacht*. It is one of the pieces in which the subject really " possessed " the composer, so that one feels throughout it a feeling of absolute conviction. There are many examples throughout of Liszt's novel and expressive use of harmony, such as this:

The seventh piece, *Eglogue*, is again prefaced with a quotation
from " Childe Harold ":

> The morn is up again, the dewy morn
> With breath all incense and with cheek all bloom;
> Laughing the cloud away with playful scorn
> And living as if earth contain'd no tomb!

Here again we have all the freshness of the country scene;
but there are also some unusual twists of harmony, particularly
in the use of unrelated 6/3 chords, which were to have an
influence on a whole generation of later composers (cf. for
instance the Flower Maiden scene from *Parsifal*):

Le mal du pays is an even more successful attempt, in that an
individual atmosphere is conjured up by very simple, folksong-
like means; here Liszt showed himself a pioneer in the field
which Grieg later explored, as we may see from this extract:

The final piece in the collection, *Les cloches de Genève*, is
dedicated to Liszt's elder daughter Blandine, who was born in
that city on 18 December 1835; it too bears a quotation from
" Childe Harold ":

I live not in myself, but I become
Portion of that around me.

Though pleasant and charming enough, it has not the originality of the rest of the set, which must remain one of Liszt's finest achievements. In it we see romantic landscape painting at its best; it was an art then being explored for the first time as far as music was concerned (a work like Beethoven's Pastoral Symphony shows a very different approach), though in a rather superficial and obvious manner; but Liszt, by the freshness of his vision, did succeed in contributing an entirely new element to the music of his time, and one which he left as a legacy to subsequent generations of composers.

With the second book of the *Années de Pèlerinage* the same tendency is continued, but in a heightened and deepened form. We have left the calm of the Swiss valleys and lakes for the sun-drenched atmosphere of Italy, with its long tradition of cultural and artistic achievement; and it is not surprising that the pieces in this second volume should all have been inspired by literary or artistic models. The list of contents is as follows:

1. Sposalizio.
2. Il Penseroso.
3. Canzonetta del Salvator Rosa.
4. Sonetto 47 del Petrarca.
5. Sonetto 104 del Petrarca.
6. Sonetto 123 del Petrarca.
7. Après une lecture du Dante, Fantasia quasi Sonata.

The inspiration for all these pieces dates from Liszt's stay in Italy with the Comtesse d'Agoult in 1838-9, and the majority of them were written at this time, at any rate in their original versions. The *Canzonetta* however was not actually put on paper till 1849; the so-called *Dante Sonata* was sketched as early as 1837 and played in this form by Liszt in Vienna in 1839; it was revised and given its present form in 1849. The three Petrarch Sonnets were published separately in 1846; the whole volume did not actually appear till 1858, and therefore is sometimes mistakenly regarded as a work of Liszt's Weimar period.

The *Sposalizio* was inspired by Raphael's famous painting, " The Marriage of the Virgin," in the Brera at Milan. It is

mainly a quiet, lyrical piece which carries still further the
tendency shown in the *Eglogue* towards the free use of notes
foreign to the main harmony, as we may see from this passage:

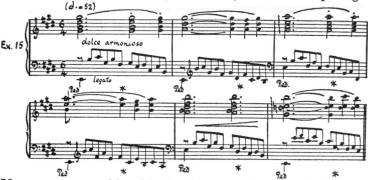

If one compares this with, for instance, the first *Arabesque* of
Debussy, it is easy to see how far Liszt was ahead of his time.
The second piece, *Il Penseroso*, is even more remarkable.
Inspired by the celebrated statue by Michelangelo in the
Medici Chapel in the church of San Lorenzo at Florence, it is
unique not only for its expression of brooding melancholy,
but also for its bold use of chromatic harmony which anticipates the style of *Tristan*—yet *Tristan* was not composed till
twenty years later! This extract from the end is typical:

To *Il Penseroso* Liszt prefaced a quotation from Michelangelo which may be roughly translated as follows: " I am thankful to sleep, and more thankful to be made of stone. So long as injustice and shame remain on earth, I count it a blessing not to see or feel; so do not wake me—speak softly! " The piece had a great personal significance for Liszt: more than twenty years later he used it as the basis of the second of the *Trois Odes Funèbres*, which, as we shall see, he intended to be his own requiem.

The *Canzonetta del Salvator Rosa* is a cheerful, unpretentious piece based on a song written by the celebrated painter of the seventeenth century. Its words are roughly as follows: " Often I change my place of being, but I shall never change my feelings; the fire of my love will remain the same, and so too will I myself." It is a good example of Liszt's simpler and more straightforward type of transcription.

The three Petrarch Sonnets exist in several versions. They were originally written in 1838-9 as songs for high tenor voice (going up to C sharp) and as such are probably the first of Liszt's songs. In this version the first sonnet is No. 104 and the second No. 47. At about the same time Liszt made a piano solo version of the songs, and it was this version which was issued separately in 1846, followed a year later by the publication of the song version. Some time before 1858 Liszt altered the piano version into the form which is now familiar to us, reversing the order of the first two sonnets at the same time, and they were then published in the complete second volume of the *Années de Pèlerinage*. About 1865 Liszt made another version of the songs, this time for much lower voice (medium tenor or baritone), and considerably simplifying them in the process; this version was published in 1883. The original tenor song version (as well as its piano transcription) shows all the romantic exuberance of Liszt's youth, as well as a strong influence of the Italian operatic *bel canto*; these songs are well worth reviving, for they are not only extremely beautiful in themselves, but also show an unfamiliar side of Liszt's art. The mature piano transcriptions are admirable examples of Liszt's power of translating music from one medium to another without any feeling of loss; taken by themselves they make excellent piano pieces, and one would certainly not have guessed that they were originally written for the voice. They are all of supreme beauty, and there is little to choose between them; many consider that No. 104 is perhaps the finest. The final song version makes a most interesting contrast to the 1838-9 songs; here the exuberance has completely disappeared, and is replaced by extreme simplicity, almost austerity. It is not perhaps surprising that Liszt's attitude to the poems should have changed with the years; but the difference between the two song versions is quite remarkable.

The final piece in the collection, *Après une Lecture du Dante*, is also the most extended and ambitious. Liszt and Mme d'Agoult often read Dante together in the 1830s; and though the title of the piece is actually taken from a poem by Victor Hugo, there is no doubt that in it Liszt expressed his own reactions to the " strange tongues, horrible cries, words of pain, tones of anger " which Dante describes in his *Inferno*. It is a strange, confused and passionate piece, perhaps incoherent and inchoate, but conjuring up a powerful and unmistakable atmosphere. It is not quite altogether satisfactory as a piano solo piece, for Liszt often seems to be trying to express things which are beyond the powers of the instrument; but a version for piano and orchestra, such as that made by Constant Lambert for the Sadler's Wells ballet *Dante Sonata*, brings out its qualities in a far clearer and more incisive manner. Liszt was to return to Dante again in his Dante Symphony; but this sonata remains an interesting and impressive attempt at the interpretation of literature in music.

As a supplement to the second volume of the *Années de Pèlerinage* Liszt wrote a set of lighter pieces entitled *Venezia e Napoli*, which exist in two versions. The first dates from about 1840, and consists of four pieces: Lento, Allegro, Andante placido and *Tarantelles napolitaines*. The first is interesting in that it is based on the same Venetian gondoliers' song which Liszt later used as the principal theme of the symphonic poem *Tasso*. Otherwise the pieces are not remarkable; the collection was engraved at the time but not printed (it may be found in the Collected Edition, II, 5), and in 1859 Liszt revised it and reissued it in the form which we know to-day. This consists of three pieces: Gondoliera, Canzone and Tarantella, of which the first and last are based respectively on the third and fourth pieces of the earlier collection. The theme of the first is described by Liszt as " La Biondina in Gondoletta, Canzone del Cavaliere Peruchini," the second is " Nessun maggior dolore, Canzone del gondoliere nel Otello di Rossini "; while the third appears to be based on themes by an otherwise forgotten composer, Guillaume Louis Cottrau (1797-1847). None of the pieces is of great moment; the best known of the three is the Tarantella, which is gay, glittering and altogether attractive and has become a very effective display piece for pianists.

The third volume of the *Années de Pèlerinage*, also mainly concerned with Italian subjects, dates from Liszt's old age, and will be considered in its appropriate place; and we may conclude this chapter by discussing the various smaller pieces, arrangements and fantasies which Liszt wrote during this prolific early period. There are not many other original works, the chief ones being the *Grande Valse di Bravura*, a brilliant salon piece, and the *Valse mélancolique*, with a charming, intimate appeal. There is also the *Rondeau fantastique sur un thème espagnol, El Contrabandista* (a song by Manuel Garcia), which is interesting both as being Liszt's first work based on a Spanish theme, and also because it suggested to George Sand (to whom it is dedicated) the idea of her story " Le Contrebandier." The piece itself, though decidedly overlong and in places trivial, has nevertheless some interesting moments. Another curious work is the *Hexameron*, which is described as " Morceau de Concert. Grandes Variations de Bravoure sur le Marche des Puritains " (*I Puritani* of Bellini). This was written in 1837 for a charity concert given at the house of Princess Belgiojoso. Six composers took part in the work: Chopin, Pixis, Thalberg, Czerny, Herz and Liszt each contributed a variation, and in addition Liszt wrote the introduction, the piano arrangement of the theme, the bridge passages between the variations and the finale. The work is interesting in that it gives a conspectus in a small space of the various styles of piano writing at the time; one can see clearly how Liszt's technique stands out from that of his contemporaries, and though Chopin's contribution is admittedly not one of his greatest works, sufficient of his individual poetical personality emerges to prove his stature. But those who would like to see the kind of music that Thalberg, Herz and Pixis actually wrote may find much of interest here.

There also date from this period two works which are rarities in Liszt's output, in that they were written for violin and piano. These are a Duo or Sonata, which remained unpublished until 1964,* and the *Grand Duo Concertant sur la Romance de M. Lafont " Le Marin "*; the latter was published in 1852, and there is a copy in the British Museum. Charles Philippe Lafont (1781-1839) was a celebrated violinist of the day, and Liszt presumably wrote the piece in the first place to play with him; he revised it in 1849 before publication. It is in the normal fantasia form of the period, consisting of Introduction, Theme, four variations and a Finale. Needless to say the piano part is just as brilliant, if not more so, than the violin part, which is sound but not exciting on the whole. There is, however, some fairly active violin writing here and there, including a short cadenza at the end of the third variation; however the piano takes its revenge with a cadenza of its own in the fourth. The Duo is chiefly interesting as a curiosity; it is noteworthy that Liszt wrote no other violin music except an *Epithalam* for the wedding of the Hungarian violinist Reményi in 1872, though in 1860 he actually began to sketch out a violin concerto for Reményi. All the other works of his which come into the category of " chamber music " were in fact written for other media, as we shall see in discussing the later works.

This was also the period of the brilliant operatic fantasies, on Halévy's *La Juive*, on the *Niobe* of the now forgotten composer Pacini—this was the work which Liszt played on the occasion of his celebrated " duel " with Thalberg—and on Donizetti's *Lucia di Lammermoor*. The latter is divided into two parts, the first being a transcription of the famous sextet, and the second based on the March and Cavatina; this division was made by the publisher against the wishes of Liszt, who thought of both parts as belonging together as one whole. Slightly later came the fantasies on Meyerbeer's *Les Huguenots* and Bellini's *I Puritani*; the latter was reissued in a revised and

---

* This work is based on Chopin's Mazurka in C sharp minor, Op. 6, No. 2; it must have been composed between 1832 and 1835.

shortened form in 1841. All these are, needless to say, brilliant virtuoso pieces, but by no means entirely negligible from the musical point of view; however exaggerated the pianistic style may often be, Liszt certainly did not write them with his tongue in his cheek, and if the result is often superficial, this merely means that these operas aroused no deep creative imagination in Liszt. Many of the later fantasies are no doubt better musically; but the arrangement of the sextet from *Lucia*, for instance, is excellently done, and does conjure up the dramatic atmosphere of that famous scene.

Apart from the operatic fantasies, Liszt also wrote two fantasias on themes from Rossini's " Soirées Musicales "—so called because Rossini composed these songs for evening parties with his friends. The fantasias are complicated virtuoso pieces of much the same type as those on operatic themes, the first being on *La Serenata and L'Orgia*, and the second on *La Pastorella dell' Alpi* and *Li Marinari*. But in 1837 Liszt transcribed twelve of the *Soirées Musicales* in a perfectly simple and straightforward manner which is much more suited to their unsophisticated character, and it is this collection which is known to-day. It contains the following pieces:

1. La Promessa.	7. La Partenza.
2. La Regata Veneziana.	8. La Pesca.
3. L'Invito.	9. La Danza.
4. La Gita in Gondola.	10. La Serenata.
5. Il Rimprovero.	11. L'Orgia.
6. La Pastorella dell' Alpi.	12. Li Marinari.

These pieces show both Rossini and Liszt at their most graceful and charming, and their unpretentiousness makes them well worthy of revival. The same may be said of the *Soirées Italiennes* on themes of Mercadante, a Neapolitan opera composer, forgotten to-day, but of whom Liszt thought highly. The titles of these are:

1. La Primavera.	4. La Serenata del Marinaro.
2. Il Galop.	5. Il Brindisi.
3. Il Pastore Svizzero.	6. La Zingarella spagnola.

A third collection consists of transcription of three songs by Donizetti, *Barcajuolo*, *L'Alito di Bice*, and *La Torre di Biasone*,

published together under the title *Nuits d'Été a Pausilippe*—
again evoking a Neapolitan atmosphere. These charming
pieces show Liszt in a far better light than some of his more
pretentious fantasies of the time, and should certainly be
rescued from oblivion.

The "partitions de piano" of this period, apart from the Berlioz
ones already mentioned, include a transcription of the *William Tell*
overture of Rossini, and—no doubt as a graceful compliment—
the vocal score of the complete opera *Esmeralda* by Louise
Bertin, daughter of the proprietor of the powerful *Journal des
Débats*, who was a protector of Berlioz. (There is a copy of
this curious work in the British Museum.) Berlioz had con-
stituted himself the " musical secretary " of Mlle Bertin—
presumably meaning that he corrected her technical faults—
and he no doubt called Liszt in to help him in this way. The
opera was produced on 14 November 1836, and was not a
success. Apart from preparing the vocal score, Liszt also made
a piano transcription of one of the arias, that of Quasimodo.
(There is a copy of this in the Paris Opera Library).

But the other transcriptions of the period were of far greater
importance, for they included two of Beethoven's symphonies
and a number of Schubert's songs. In 1837 Liszt transcribed
Beethoven's 5th and 6th Symphonies and possibly also the 7th;
this was followed in 1841 by the Marcia Funebre from the
Eroica. About 1850 the publishers, Breitkopf and Härtel,
began to press Liszt to complete the whole series; but he was
not ready to undertake this task till 1863, when he transcribed
the remaining symphonies and revised the earlier transcriptions. The last movement of the 9th Symphony caused
him considerable trouble, and at first he considered it impossible
to arrange for piano solo; but eventually this task too was
completed, and the whole series was published in 1865. At
the time when Liszt made his first transcriptions of the
symphonies, Beethoven was by no means popular in all circles
in Paris, and Liszt was in fact doing the same kind of pioneering
work for him that he had already undertaken for Berlioz.
The transcriptions, while in general keeping to the simplicity
and directness of the originals, translate orchestral effects into
pianistic ones much as in the arrangement of Berlioz' Fantastic
Symphony; they are published in the Collected Edition

(Arrangements, Vols. II and III), and the reader may see for himself with what care and fidelity Liszt has reproduced the spirit, and as far as possible, the letter of the originals. Liszt's first Schubert transcription, *Die Rose* (*Heidenröslein*), dates from 1833; it was followed five years later by a whole group which included *Lob der Tränen* and *Der Gondelfahrer*, a collection of twelve songs containing, among others, *Erlkönig*, *Gretchen am Spinnrade*, *Ständchen* (" Hark, hark, the lark ") and *Der Wanderer*, and finally in 1839 by the *Schwanengesang* and *Winterreise* complete. The tale of Liszt's Schubert transcriptions will be continued in our next chapter; but we may already see the great devotion which Liszt had to the earlier composer, who, like Beethoven, was insufficiently appreciated outside Vienna at that time, and certainly needed Liszt's championship. The transcriptions catch the very essence of Schubert, a composer with whom Liszt had a great deal in common; they both poured out their riches in all directions, perhaps too prolifically and with too little thought at times—but they both had the feeling of the divinity of music, of that innate and natural beauty which cannot be acquired merely by taking pains; and if Schubert's songs are sometimes dull and Liszt's piano pieces sometimes superficial, we can forgive them both for the complete spontaneity of their approach. Liszt's transcriptions of Schubert may be overdone at times; but they do not contradict the essential freshness of Schubert's genius. And, as we shall see in connection with the *Wanderer Fantasy*, Liszt not only shared many of Schubert's moods and methods of expression, but was also considerably influenced by his attempts to evolve new musical forms.

We may complete our study of Liszt's early period with a reference to yet another virtuoso piece, the *Grand Galop Chromatique* of 1838. This work is the essence of all concert-platform fireworks; it is the grand finale *par excellence*, and we can well imagine Liszt ending his concerts with it, with fevered gestures and hair flying in all directions. Nevertheless it is by no means negligible as a piece of light music, which is all that it is intended to be; it is short, well shaped, entertaining and has good tunes, and there is really no reason why a serious composer should not unbend at times if he does it as well as this. We shall see more of this side of Liszt during the next period, his years as a travelling virtuoso; but meanwhile we may recall the remarkable variety and richness of the works we have discussed so far, ranging as they do from the romantic expressionism of the *Harmonies poétiques et religieuses*, through the virtuosity of the Transcendental and Paganini Studies, to the poetical imagery of the *Années de Pèlerinage* and the dazzling glitter of the salon pieces. It was surely no mean feat to have created a corpus of works of such variegated distinction by the age of twenty-eight.

CHAPTER II

THE VIRTUOSO PERIOD (1839-47)

THE NEXT eight years of Liszt's life were spent by him in constant travel and concert-giving, ranging from Dublin to Moscow, and from Lisbon to Constantinople. It is not surprising therefore that no work of really major importance was composed in this period; what is surprising is, firstly, the number of works which Liszt actually did find time to write, and, secondly, the fact that this period proved in the end to be one of incubation for the major works of the Weimar period; for many of the latter were sketched out, or at any rate thought of in some form, during Liszt's travels and only given their final shape in the comparatively stable atmosphere of Weimar.

Up to now, as we have seen, Liszt had written comparatively little for other media except the piano; but from this time on he began to branch out in other directions, and in the remaining chapters of this book it will be best to group his works according to the resources for which they were written. His piano music still remains the major part of his output in this period, of course, and in the main continues the tendencies seen towards the end of the last chapter. There is one further study, originally written in 1840 under the title *Morceau de Salon, Étude de Perfectionnement* for the " Méthode des Méthodes de piano " of Fétis and Moscheles; this piece was revised and republished in 1852 with the title *Ab Irato*, by which it is known to-day. It is a short and mainly violent piece, attractive and effective of its kind; towards the end there is a calmer section in which one of the main themes of *Les Préludes* is clearly foreshadowed.

The works in dance forms include an excellent Galop in A minor, just as brilliant as and musically superior to the *Grand Chromatic Galop*; for some reason this piece was never published in Liszt's lifetime, and first appeared in the Collected Edition in 1928. (An orchestration of it by Gordon Jacob was used in the Ball Scene of the Sadler's Wells ballet *Apparitions*, and recordings of this version have been made by Constant Lambert and Robert Loving.) It has all the gaiety of Offenbach at his best, and deserves to be played, though its technical difficulties are formidable. There is also an admirable waltz on themes from Donizetti's *Lucia* and *Parisina*, which has all the freshness and brilliance of Liszt at his youthful best; this too is a piece which is well worth reviving. The other original piano works are mainly short salon or album pieces; but there are a number of important fantasies and transcriptions. The fantasies include those on the Tarantella from Auber's *Muette de Portici*, on Bellini's *Sonnambula* and *Norma*, Donizetti's *Lucrezia Borgia* (in two parts) and the Funeral March from *Dom Sebastien*, Mozart's *Don Giovanni* and Meyerbeer's *Robert the Devil*. (A further fantasy, on Mozart's *Figaro*, was left unfinished, though actually played by Liszt in Berlin in 1843; it was completed and published by Busoni in 1912.) These in some ways represent the high point of Liszt's work of this type; whatever one may think about the operatic fantasy as a musical form, there is no doubt that in many of these works Liszt completely transcended his original material and produced a kind of re-treation of the thoughts of the composers which raises them to a far higher musical level. Mozartian purists may argue about Liszt's approach to *Don Giovanni* and *Figaro*; but there is no doubt that the fantasy on *Norma*, for instance, is far more than a mere effective concert piece, and gives a summary in short space and concentrated form of the whole musical content of Bellini's opera; further than that, as Kaikhosru Sorabji wrote: " Bellini's themes never had, by themselves, the grandeur and magnificence that Liszt is able to infuse into them." The same is true, perhaps to lesser extent, of the fantasies on *Sonnambula*, *Dom Sebastien* and the *Muette de Portici*; before dismissing these works as mere salon fireworks one should note the extraordinary breadth and power which Liszt was able to impart to material which in itself was often somewhat undistinguished.

The *Don Giovanni* fantasy represents a different problem, in that here Liszt was dealing with an undisputed masterpiece, and it might therefore seem impertinent of him to attempt to add anything to it. Of course the nineteenth-century attitude to such things was not as purist as ours is, and Liszt would have been quite justified in the eyes of his contemporaries in giving his own interpretation of even such a great work as this. And in fact one cannot say that Liszt has vulgarised it; his approach is certainly broader and more romantic than is in fashion to-day, but he was interested in the work as a human and dramatic story, and it was this aspect of the opera which he wanted particularly to bring out. He did this by taking three scenes which typify three different sides of the drama: a slow introduction based on the Statue Music when the Commendatore appears at the supper scene, a middle section with two variations on the duet between Don Giovanni and Zerlina, and a finale on Don Giovanni's " Champagne Aria." These three episodes represent the essence, though not the whole, of the story; at any rate one is presented with the three main ideas of justice, seduction and carefree enjoyment which form the mainspring of the opera. Liszt takes each idea as the basis of a musical section and works it out so as to represent the dramatic development of that particular scene in the opera; but the three ideas are each treated separately, and there is no question of a pictorial " symphonic poem " representing the actual course of the action. (The only exception is a passage towards the end of the finale where the theme of the Commendatore reappears to cast a blight on Don Giovanni's gaiety.) Personally I find the work a completely satisfying interpretation by one composer of the ideas of another; it, of course, cannot make the same effect as a hearing of Mozart's opera, but it is not intended to—it is Mozart-Liszt, and not Mozart, and one should appreciate it for what it is. And Bernard Shaw's opinion is worth quoting: " When you hear the terrible progressions of the statue's invitation suddenly echoing through the harmonies accompanying Juan's seductive ' Andiam, andiam, mio bene,' you cannot help accepting it as a stroke of genius—that is, if you know your *Don Giovanni au fond*."

The *Figaro* fantasy is a slighter and simpler work; it consists of an introduction in which Figaro's " Non più andrai " is heard as it were at a distance and in a visionary manner; then an ingenious twist leads to a full treatment of Cherubino's " Voi che sapete," beginning simply, but accumulating complications as it proceeds; in the final section " Non più andrai " returns and brings the work to a brilliant end. The work had been very nearly completed by Liszt, and Busoni's additions do not amount to a great deal. The *Figaro* fantasy is much more in the manner of Liszt's normal operatic fantasies than that on *Don Giovanni*, and one cannot say that it presents any real interpretation of Mozart's opera.

The remainder of Liszt's piano works during this period were based on national themes of various countries. These took many and varied forms; some were full-length fantasias, like those on *God Save the Queen* and on Spanish themes, written as tributes to their respective countries on the occasion of Liszt's tours there; others were comparatively short and simple, such as the arrangement of two Pyrenean folk-songs, *Faribolo Pastour* and *Chanson du Béarn*, written on the occasion of his meeting, the first for many years, with Caroline de Saint-Cricq, now Madame d'Artigaux. The Spanish fantasia, which has one theme in common with the later (and far better) *Rhapsodie Espagnole*, belongs to the category of Liszt's over-elaborate and over-written works; and in fact it was not published until after his death. The Pyrenean melodies are charming; and so are the two " Mélodies russes, Arabesques," of which the first is a transcription of Alabieff's song " The Nightingale," and the second a gay and brilliant " Chanson Bohémienne." Apart from the Hungarian works, to which we shall return in a moment, the remaining folk-song transcriptions of this period are of no great importance—with the exception of the last of them, the *Glanes de Woronince*, written in 1847-8 on gipsy themes which Liszt heard on the estate of Princess Sayn-Wittgenstein in the Ukraine. The collection consists of three pieces: *Ballade d'Ukraine*, *Mélodies polonaises* and *Complainte*, the first and last being in the form of Dumkas, while the second contains the theme of Chopin's " Maiden's Wish," which Liszt later included in his set of six transcriptions of Chopin's Polish songs. " Glanes " means " gleanings "; and the whole collection reflects the immediate impact of Liszt's meeting with the woman who was to change the subsequent course of his life.

These years were also the period when Liszt began to take an active interest in Hungarian popular music—an interest which remained with him till the end of his life. In 1840 he wrote a Heroic March in the Hungarian style,* a fine, compact and exciting piece which was later expanded—not to its advantage—to become the basis of the symphonic poem Hungaria. This march, which clearly merits revival, was followed four years later by a Hungarian Storm March, a less distinguished piece which Liszt revised and orchestrated in later years. But the most interesting Hungarian works of this period were the pieces which were eventually transformed into the Hungarian Rhapsodies—the works by which Liszt's name is still chiefly known to the general public.

As may be seen from the catalogue at the end of this book (Nos. 242-4), the history of these works is somewhat complex. Liszt had of course been familiar with popular Hungarian music, and in particular with that of the gipsies, in his boyhood, and he heard it again on his return to Hungary during his concert tours. He was seized with the idea of transcribing this music for the piano, at first in the form of improvisations based on the improvisatory methods of the gipsies themselves— he performed a fantasia of this type on the Rákóczy March in Pest in 1839—and later as written and published works. Between 1840 and 1847 he published seventeen pieces in ten volumes on these themes, under the titles *Magyar Dallok* and *Magyar Rhapsodiák*, as well as a separate collection of three pieces on some of the same themes called *Ungarische National-melodien*. In addition the Liszt Museum possesses four further unpublished pieces of the same type; so that when Liszt came to write the Hungarian Rhapsodies themselves he was able to use this material as the basis for all but three (Nos. 1, 2 and 9) of the first fifteen Rhapsodies. (Rhapsodies Nos. 16 to 19 belong to the last years of his life and are discussed later.)

The whole question of Liszt's relation to Hungarian gipsy music is a large and complicated one which needs more space than it is possible to give it here (those interested will find further details in Zoltán Gárdonyi's " Die ungarischen Stilei-gentümlichkeiten in den musikalischen Werken Franz Liszts," Berlin and Leipzig 1931). Liszt himself wrote a book on " The Gipsies and their Music in Hungary " which sets forth his

* Two further pieces in this style (No. 693 in catalogue) have recently been published by the Liszt Society.

attitude; it should be said that he tended to over-estimate the part played by the gipsies at the expense of the older native Hungarian folk music, and his successors like Bartók and Kodály have reversed this emphasis. Liszt of course admired the romantic elements in the gipsies' style and their improvisatory effects; but he did not realise that the characteristic Hungarian melody is the product of several successive melodic methods, of which the instrumental *fioriture* of the tziganes who come from Asia is only one. Even so Liszt made most use, not of the traditional anonymous tzigane melodies, but of themes by various dilettante Hungarian composers who were quite well known by name, and who were not themselves tziganes—he merely applied to these the tzigane style of ornamentation and transcribed them for piano. However he was the first to make a modern piano transcription of the so-called " Verbunkos " type of Hungarian folk music, which was a characteristic feature of it from the beginning of the nineteenth century onwards—and if the Rhapsodies may sound cosmopolitan to a true Hungarian, they nevertheless represent a considerable achievement in the art of folk-song transcription.

The characteristic instruments of the tziganes were the solo violin, often accompanied by other strings, the clarinet, and the cymbalom; there were frequent improvisatory and ornamental passages and cadenzas, and in transcribing this music Liszt had to take account of the natural style and colour of these instruments, which meant in practice reproducing violin and cymbalom effects on the piano. This he did in general with great success, as many passages in the Rhapsodies show. But on the whole the Hungarian Rhapsodies do not rank among Liszt's best works—the thought is often too limited and conventional, and there is too much striving for superficial brilliance and effect. Perhaps the best summing up is that made by one great Hungarian composer on another, by Béla Bartók in an article written on the occasion of the centenary of Liszt's birth in 1911. " The Hungarian Rhapsodies," he wrote, " which should say the most to us, are his least successful works (perhaps that is why they are so generally known and admired). Alongside strokes of genius we find altogether too conventional ideas—gipsy music, sometimes mixed up with Italianisms (No. 6), sometimes in complete formal confusion (No. 12)."

One small point remains to be mentioned, concerning the so-called " 20th Rhapsody." This was edited and performed by Busoni, and was published in 1936 as "Roumanian Rhapsody," edited by Dr. Octavian Beu. This work is in fact the twentieth of the *Magyar Rhapsodiák*, not of the Hungarian Rhapsodies themselves; it is partly based on themes heard by Liszt on his travels through the Roumanian principalities in 1846-7—these were typical Batutas and Horas of Moldavia and Transylvania. But not all the themes in it are Roumanian, and parts of it were later incorporated in the 6th and 12th Rhapsodies.

Six of the Rhapsodies (Nos. 14, 12, 6, 2, 5 and 9) were later issued in an orchestral version, described as " arranged by the composer and Franz Doppler "; but according to Liszt's English pupil and friend Walter Bache, " Doppler was a flute player who did arrange some of these Rhapsodies for orchestra. So when Liszt published the set of six, he very generously put Doppler's name on the title out of compliment to him; but Doppler had nothing to do with them. If you like to mention this characteristic of Liszt's kindness, do so; but don't mention my authority for it, which is Liszt himself." As we shall see in discussing the orchestral works, Liszt ceased employing collaborators in his orchestral scoring from about 1854 onwards, and thereafter wrote out all his own scores himself; in any case the orchestral versions of the Rhapsodies are extremely effectively done, and deservedly hold their place in the popular repertoire. The 15th Rhapsody (Rákóczy March), which exists in at least three different piano versions, was also scored for orchestra later in Liszt's life; Liszt somewhat naturally wished to avoid competition with Berlioz' brilliant orchestral arrangement of the same theme, and therefore withheld his own version for many years. Though certainly effective, it does not generate the enormous excitement of Berlioz' version.*

It only remains now to discuss the piano transcriptions of this period. In general these continued on the same lines as those of the late 1830s; there are transcriptions of Beethoven's

---

* Liszt's earlier piano arrangements of the Rákóczy March (and probably some form of his orchestral arrangement) were actually written *before* Berlioz' version (1845); in fact, it was Liszt who first introduced Berlioz to the march; see Liszt's Briefe II, 336.

Gellert Songs, Op. 48 and Septet, and of Schubert's *Müller-lieder* and other songs, as well as of the *Divertissement à l'hongroise* and some of the piano duet marches. In transcribing for one player what Schubert had written for two, Liszt often made formidable demands of the pianist, who was expected to play rapid passages like this:

Ex. 17

But, though somewhat over-elaborated here and there, these transcriptions do remain faithful to the spirit of their originals. A number of new composers were also added to the list of those transcribed; these included Mendelssohn (a group of songs), Weber (Freischütz, Oberon and Jubilee overtures, as well as two songs from *Leyer und Schwert* and a fantasy on Freischütz), and Glinka (Tscherkessenmarsch from his newly produced opera *Russlan and Ludmila*); this was Liszt's first real contact with the Russian nationalist school, which was to assume such importance in his later years. In addition Liszt began to transcribe some organ preludes and fugues of Bach— one of the fairly rare cases of his taking an interest in music which was not by his contemporaries. Bach was of course not very much in fashion at that time, and Mendelssohn was struggling heroically to put him on the map; so Liszt's transcriptions may be regarded to a great extent as a pioneering effort. They are done in a perfectly simple and straightforward manner, and are the forerunners of the many transcriptions of Bach's organ works for the piano made by a number of modern composers from Busoni downwards.

Liszt's other transcriptions of this period are of no great musical value, being mostly pieces of politeness written on themes by his aristocratic friends; and we pass on to an extremely remarkable work which would appear on stylistic evidence to have been written at this time (if not actually earlier)—the so-called " Malediction " concerto for piano and strings. This work has a curious history. In 1827, when Liszt was sixteen, he played in London a concerto which Moscheles described as containing " chaotic beauties," but of which all trace later seemed to have disappeared. But in the Weimar Liszt Museum there was afterwards discovered sixteen pages of sketches for a concerto for piano and strings which clearly originated in Liszt's youthful years, and which contained three themes which were later used in the *Malediction*. The *Malediction* concerto itself was not published until thirty years after Liszt's death; the MS. of this too was found at Weimar. It seems clear that during the 1830s Liszt was obsessed with the idea of writing one or more piano concertos; the first theme of the E flat Concerto is found in a sketch book which dates from the early 'thirties, and the first version of the A major Concerto dates from 1839. But Liszt's lack of technical knowledge of orchestration delayed these projects until the Weimar period; and it seems more than likely that he had decided first to try out his hand by writing a work for piano and strings only, based on his youthful concerto. The Weimar MS. of the *Malediction* is written in a hand different from those of all the copyists who worked for Liszt from 1847 onwards; it seems reasonable therefore to date it early in the 1840s, if not even earlier.

I have said " so-called *Malediction* concerto " above, for in fact Liszt left the work untitled. He did however make additions and corrections to the copyist's MS., and over some of the principal themes he wrote phrases indicative of their character. Thus " Malediction " only applies to the opening theme, a startling phrase, in some ways similar to the opening of *Orage* from the Swiss book of the *Années de Pèlerinage*; it is found in a sketch book which dates from the early 1830s.

Ex. 18

After a short piano cadenza, mainly based on the clash of two chords a tritone apart—an effect not afterwards paralleled till *Petroushka**—the second main theme, marked "orgueil," enters on the piano.

Ex. 19

This theme was later used in the finale (Mephistopheles) of the Faust Symphony, where it is the only one which is not a parody of those of the first movement. A later theme, a romantic 'cello solo with piano accompaniment (p. 7 of the Breitkopf score) is marked "pleurs—angoisses—rêves," the latter word being later crossed out and "songes" substituted; and a brilliant *Vivo* passage (p. 11) is marked "raillerie." The concerto is thus a succession of mood pictures, poetical, romantic and emotional, and extremely characteristic of Liszt at this period—he was even thinking of introducing into it a transcription of Schubert's "Du bist die Ruh" at one point, but afterwards, wisely, thought better of it. It is admittedly a patchy work, and some of the string writing is extremely awkward for the players—though curiously enough its actual sound in performance is perfectly effective; but it contains a number of extremely interesting things, and is well worth performing by those who can cope with the difficult piano part.

This period is also important in that it saw the beginning of Liszt's vocal works. These fall into two categories—choral works (mainly secular at this time) and solo songs. The choral works are not of great importance; there are a number of pieces for male chorus, a medium which Liszt chose both because it was comparatively easy to master, and because he liked it as an actual means of expression. Many of these have remained unpublished, and even those which have appeared in print cannot be said to add much to Liszt's reputation—a list will be found in the catalogue at the end of the book. A more important work was the First Beethoven Cantata, written for the unveiling of the Beethoven monument

* Incidentally the parallel passage in *Petroushka* is in fact called "Malédictions de Petroushka"—a curious coincidence, as Liszt's *Malediction* was still unpublished at the time Stravinsky wrote the ballet.

in Bonn in 1845, and performed twice in succession under Liszt on that occasion. The text by O. L. B. Wolff is not particularly inspiring, and the music is hardly more than competent; it was however Liszt's first large-scale work for chorus and orchestra, and does show an efficient handling of the medium. Berlioz, who was present at the first performance, wrote a glowing account of the work; but posterity in general has not agreed with him. The only other choral work of this period which deserves mention is the *Ave Maria I* (1845) for mixed chorus—the first of the long succession of religious choral pieces which occupied Liszt up to the end of his life. With a few exceptions, which we shall see in a later chapter, Liszt was not at his most characteristic as a choral composer; but he did know which parts of the vocal compass sounded best, and his knowledge of Italian opera ensured that his vocal writing, even if not in the normal choral tradition, would at least be singable and effective.

Some of these choral works have connections with other and better-known works; for instance, *Les quatre Élémens* provided the thematic material for *Les Préludes*, which was originally designed as an introduction to it: " Über allen Gipfeln ist Ruh " was later transformed into the well-known solo song, and the *Arbeiterchor* reappears in an altered form as the march in the symphonic poem *Mazeppa*.* The First Beethoven Cantata also has a link with the second cantata of 1870, in that both contain an orchestration of the Adagio from Beethoven's Trio in B flat, Op. 97 (The Archduke). But it is noteworthy that after this period Liszt's secular choral works became few and far between, and his attention became increasingly turned towards sacred music.

The songs are another matter, for they represent the first exploitation of a fruitful vein which Liszt continued to explore for the rest of his life. Of his seventy songs, about thirty were written during this period (apart from the first versions of the three Petrarch sonnets, which were discussed in our last chapter). Many were revised and reissued later, but in essence their inspiration dates from this time. The first is a charming strophic song, " Angiolin dal biondo crin," written in 1839

---

* Three of the collection called " Für Männergesang " were also published for piano under the title " Geharnischte Lieder " (Songs in Armour).

in honour of Liszt's daughter Blandine. Then followed a
number of French and German songs, mainly settings of
Goethe, Heine and Victor Hugo, in addition to three songs
from Schiller's *Wilhelm Tell* and settings of several minor poets.
It is interesting to compare the earlier with the later versions
of many of these songs; practically all the available versions
are published in the three volumes of the Breitkopf Collected
Edition, the earlier versions in Vol. 1, and the later in Vols.
2 and 3. In most cases the later versions, which are the ones
usually known and performed to-day, represent a considerable
improvement. In the 1840s Liszt had certain disadvantages
as a song-writer; he was a virtuoso pianist who tended to write
over-elaborate accompaniments; he was steeped in the feeling
of Italian opera, and therefore was inclined to over-dramatise
the most simple lyrical poems; and he was as yet insufficiently
at home with German traditions to avoid making mistakes in
setting German words. For instance, in his setting of the
famous Mignon song, the opening phrase is throughout
accentuated as " Kennst *du* das Land? " which no German
would dream of saying. In these early settings he also misunderstood to some extent the feeling implicit in the German
words; for instance the first version of " Der du von dem
Himmel bist " rises to a violent emotional climax which is
quite out of keeping with the spirit of the words; and often he
let the musical thought have too much control of the setting—
for instance, at the end of the exquisite lyric " Über allen
Gipfeln ist Ruh " the constant repetition of the last two lines
" Warte nur, balde Ruhest du auch " produces an exaggerated
effect. Another example, though a more successful one
musically, is Liszt's setting of the *Lorelei*. The poem has the
quality of a folk tale, and the well-known strophic setting of
Silcher, with its straightforward tune, provides a treatment
which is at any rate adequate for the words. Liszt however
treats the poem in a more elaborate manner; an introduction,
in the style of a recitative (Ich weiss nicht was soll es bedeuten)
leads to a descriptive setting of the story, culminating with a
graphic picture of the boatwreck in the accompaniment; the
last section consists of a setting of the last line of the poem
(Und das hat mit ihrem Singen die Lorelei getan), at first as a
recitative like the introduction, and then repeated several

times over the music which has previously represented the Lorelei. Musically the song makes a good whole; but some violence is done to the poem. Similarly in the first setting of " Es war ein König in Thule " there is an elaborate piano accompaniment representing the sinking of the goblet beneath the waves—whereas in Goethe's Faust the song is an old ballad sung by Gretchen to herself quite quietly and introspectively. Peter Raabe, in his " Franz Liszt " (Vol. II, chapter 4), has given an interesting comparison of the different versions of Heine's " Im Rhein "; in the first the accompaniment is dominated throughout by figures portraying the flowing of the river, whereas in the second the real point of the poem, the recognition by the poet of the features of his beloved in the picture of the Madonna, is emphasised by the disappearance of the rushing accompaniment and the complete change in the character of the music at these words. Liszt's sensitivity to lyrical poetry greatly matured during the Weimar years, and in practically every case these revised versions of the songs are great improvements on the originals.

But in spite of these criticisms, Liszt remains a very much underrated song-writer. He had a very genuine pictorial and lyrical gift, and he saw to it that in his songs the voice and piano parts were integrated into a whole—there is no question of an all-important melody with a conventional accompaniment, as in the songs of some of his contemporaries. He was at first more at home in his settings of French words, and his group of Victor Hugo songs contains a little masterpiece, " Comment, disaient-ils," the beautiful " O quand je dors " and the delightful " S'il est un charmant gazon," as well as three or four others of great charm. Of the Goethe settings " Freudvoll und leidvoll " has become well known, but " Der du von dem Himmel bist " in its later version, " Wer nie sein Brot mit Tränen ass " (of which Liszt made two different settings), and " Mignons Lied " (in spite of the defect mentioned above) are all among Liszt's finest songs. So are the three songs from Schiller's *Wilhelm Tell*, which give a feeling of Alpine freshness comparable to the Swiss book of the *Années de Pèlerinage*. But of the major German poets it was Heine whom Liszt interpreted with the greatest feeling and subtlety; apart from the settings mentioned above, " Du bist wie eine

Blume " has all the simplicity and charm of the original, and
" Vergiftet sind meine Lieder " has a remarkable dramatic
power and some extraordinary harmonic effects, especially
for a song written as early as 1842; these two passages, one
from the beginning and one from the climax of the song, may
exemplify this.

Of the other songs of this period, " Die Vätergruft " deserves
mention for the stark simplicity and effectiveness of its opening
and final sections (incidentally the orchestration of this song,
made for his London visit in 1886, was the last work that
Liszt completed). " Was Liebe sei " and " Wo weilt er? "
are both charming and in lighter vein; " Ich möchte hingehn,"
though overlong and hardly satisfactory in itself, is interesting
both for its connection with Caroline de Saint-Cricq, and
because it contains this phrase, written more than ten years
before Wagner wrote the first note of *Tristan*:

The resemblance is no doubt accidental; and a similar theme,
as we shall see, is the basis of the second " Aux Cyprès de la
Villa d'Este," written in 1877. The setting of Dumas' " Jeanne
d'Arc au bûcher " almost amounts to a dramatic scena, and,
as we shall see in the next chapter, may have some connection
with Liszt's operatic plans, which were maturing in this
period. " Isten veled " is one of Liszt's rare songs in the

Hungarian style, and in this case is an actual setting of a Hungarian poem, though Liszt's knowledge of his native language was extremely limited. But the melody which above all others has made Liszt's name universally famous also belongs to this period; in 1847 he published a song called " O lieb, so lang du lieben kannst," which is now known all over the world as *the* Liebestraum. As a matter of fact there are three Liebesträume: Nos. 1 and 2, " Hohe Liebe " and " Gestorben war ich," were songs written about 1849, and the three were transcribed for piano and published as " Liebesträume, 3 Notturnos " in 1850. Little more need be said, except that Nos. 1 and 2 are just as beautiful, if not more so, than their famous sister, and all three deserve to be performed more often in their original song form.

The tale of Liszt's piano compositions, choral works and songs will be continued in our next chapter, together with the important additional categories of works for orchestra and for organ; and it will then become clear that, apart from the works discussed in this chapter, many larger-scale works were maturing in Liszt's mind during his years of travel as a virtuoso. Though Liszt undoubtedly wasted a great deal of time during this period on brilliant trifles, his progress as a composer did continue to show a steady development, and culminated in the full-scale flowering of the Weimar years.

CHAPTER III

THE WEIMAR YEARS (1848-61)

THE WEIMAR PERIOD was that of Liszt's greatest productivity; music poured from his pen, and his output is all the more astonishing in that not only do many of the works show extremely novel and original characteristics, but also because at the same time Liszt was constantly occupied with productions of important and difficult new works on the stage of the Weimar theatre—not to mention the numerous essays on musical subjects which he wrote at the same time. Admittedly he was helped in his literary work by the Princess Sayn-Wittgenstein, and his pupils took some of the weight of routine matters off his shoulders; but even so a period which produced the first twelve symphonic poems, the Faust and Dante Symphonies, a number of major piano works, including the Sonata, as well as revisions of his earlier piano pieces and songs—not to mention numerous transcriptions and some larger-scale vocal works—must indeed be considered a remarkable one. Inevitably there were drawbacks to this activity; many of the works of this period give the impression of having been written in too much of a hurry for the thought contained in them to have really crystallised, and although Liszt frequently and tirelessly revised his works, both before and after performance and publication, there was a certain element of hit-or-miss in his compositional processes which tended to produce uneven results.

The first important piano works on which he worked during these years was the collection called *Harmonies Poétiques et Religieuses*, written at various times between 1847 and 1852. This consists of the following pieces:

1. Invocation.
2. Ave Maria.
3. Bénédiction de Dieu dans la solitude.
4. Pensée des morts.
5. Pater noster.

6. Hymne de l'enfant à son reveil.
7. Funérailles, October 1849.
8. Miserere, d'après Palestrina.
9. Andante lagrimoso.
10. Cantique d'amour.

Few of these are actually new compositions: Nos. 2, 5 and 6 are transcriptions of choral works written about 1846, No. 4 (as we have seen) is a revised version of the single piece *Harmonies Poétiques et Religieuses* (1834) which was discussed in Chapter 1, and No. 8, as its title implies, is a transcription of Palestrina. But the collection evidently had a strong personal significance for Liszt; it is dedicated to Princess Sayn-Wittgenstein, and it was the work which he took the greatest pleasure in playing to his friends in later years, when he had retired from public activity. Musically the collection is extremely uneven; the Ave Maria, Pater noster and Miserere are more or less straightforward arrangements of liturgical music of no particular importance. Lamartine's preface to his collection of poems (cf. p. 11) is now prefixed to the set as a whole, and in addition there are quotations from poems in this collection at the head of Nos. 1, 3 and 9; No. 6 was of course originally a setting of Lamartine's words too. The Invocation is somewhat fussy and overdone; *Pensée des Morts* is weaker and more conventional than its original version, and Nos. 6 and 9 give a curiously negative effect. But *Cantique d'Amour* is a fine expressive piece, and the *Benediction of God in Solitude* and *Funérailles* must both be reckoned among Liszt's masterpieces.

The *Benediction* is indeed almost unique among Liszt's works in that it expresses that feeling of mystical contemplation which Beethoven attained in his last period, but which is rarely found elsewhere in music. Lamartine's poem begins:

> D'où me vient, O mon Dieu, cette paix qui m' inonde?
> D'où me vient cette foi dont mon coeur surabonde?

and Liszt has truly expressed that atmosphere of peace and faith which the words suggest. The touching simplicity of the final passage shows that Liszt, like Beethoven, could express the most sublime thoughts in completely unadorned language when the mood was upon him, and the *Benediction* is rightly acclaimed as one of his finest and most original works.

Funérailles, though equally fine in many ways, portrays a mood which we find frequently elsewhere with Liszt—that of the heroic elegy. Its date, October 1849, refers not, as has sometimes been thought, to the death of Chopin, but to those of Prince Felix Lichnowsky and Counts Ladislaus Teleki and Lajos Batthanyi; of these Batthanyi was a victim of the Hungarian revolution of October 1849. It is certainly a most dramatic piece, and the middle section with its trumpet calls and galloping octaves makes a very exciting effect; it is one of Liszt's best works in this genre.

About 1848 Liszt wrote three further studies, which were published in the following year as *Trois Études de Concert*—though in one edition they appeared as " Trois Caprices Poétiques—Il lamento, La leggierezza, Un sospiro," and these titles are still sometimes found to-day. The keys are A flat, F minor and D flat, and the studies have become reasonably well known, particularly the last one, which is a beautiful poetical piece. The influence of Chopin may be discerned in the other two, but in general they are longer, more elaborate and less concise than Chopin's own studies. At this time, however, Liszt did write quite a number of works in the forms immortalised by Chopin—two Ballades, two Polonaises, a Berceuse in two different versions, a Mazurka Brillante and some revised versions of his earlier waltzes. These works are uneven in quality; the Ballades are very episodic, though they contain a number of fine things, especially the second one in B minor, which Mr. Sacheverell Sitwell has rightly described as being " concerned, as it were, less with personal suffering [in contrast with Chopin's Ballades] than with great happenings on the epical scale, barbarian invasions, cities in flames—tragedies of public, more than private, import." The opening part of this Ballade exemplifies a device of which Liszt made much use during this period—the first section, in B minor, is immediately afterwards repeated note for note in B flat minor. This type of repetition gave Liszt a bad name and a reputation for laziness; though admittedly it is not a very happy innovation, it was one which was bound up with his search for new forms. This question will be discussed later in connection with some of Liszt's larger works; but clearly in this case Liszt was aiming at a varied form of the repeat marks to be found at the end of the exposition of a classical sonata; while he felt that from the formal point of view the section required repetition, the fact the repeat took place in a different key would add extra variety.

The two Polonaises, though undeniably effective, tend to be overlong and bombastic, and lack the subtlety of Chopin's works in this form; and the most interesting pieces in this group are the two versions of the Berceuse. The first version, written in 1854, is clearly modelled on Chopin's Berceuse; it uses the same key, D flat, and the same idea of a consistent tonic pedal throughout. Nevertheless it can well stand on its feet as a charming piece in which Liszt most successfully exploits a very simple method of writing. The second version, written eight years later, provides a most startling contrast; for while the basic figuration and harmonies remain the same, the melodic line is embellished with the most complicated ornamentation imaginable, and though the piece is not exactly " spoilt " by this, it is certainly presented in a fantastically different aspect.

Apart from these and other smaller pieces, three of Liszt's biggest piano works date from this period. These are the *Grosses Konzertsolo* (1849), the Scherzo and March (1851), and the Sonata (1852-3). The *Grosses Konzertsolo*, written for a piano competition at the Paris Conservatoire, has several points of interest, though it is not in itself one of Liszt's greatest works. It anticipates the Sonata in two ways: firstly by the use of a theme, which in varied forms is the mainspring of both the Sonata and the Faust Symphony:

Ex. 22

Secondly, it is written in the three-movements-in-one form which Liszt used so successfully in the Sonata, and which has provided a model for so many later composers. Indeed this work may almost be said to show the genesis of this form, for the original version, which is preserved in MS. in the Weimar Liszt Museum, does not contain the *Andante sostenuto* middle section, which was an afterthought on Liszt's part. His earlier large-scale works, such as *Vallée d'Obermann* and the *Dante*

Sonata, were usually written in the form of one loosely constructed movement, which might contain contrasting episodes, but otherwise consisted entirely of varied treatments of one or two main themes. In the *Bénédiction de Dieu* Liszt did write a contrasting middle section, based on an entirely new theme, which returns at the end of the work after the repeat of the opening section; but here there is little variation in speed or even mood between the two sections. The *Grosses Konzertsolo* was evidently originally intended to follow the form of the *Dante Sonata*; but by adding a middle section, no doubt for the sake of contrast in the first place, Liszt was able to create an entirely new form. In its full state of development this consisted of a rearrangement of the normal elements of the classical sonata form, into a shape of the following kind:

First section: Exposition and development.
Second section: Slow movement.
Third section: Further development of first section and reprise.
Fourth section: Coda (including part repeat of second section).

This type of integrated form had an enormous influence on the music of the second half of the nineteenth century; it also survived into the twentieth, for instance in the English fantasy form and in some of Schoenberg's earlier works, like the first string quartet and first Chamber Symphony, which use similar methods; and even Sibelius' Seventh Symphony owes something to it, though here the slow sections act as a framework to the Allegro and Scherzo.

The *Grosses Konzertsolo* of course only shows this type of form in embryo, and further discussion of it must be deferred till slightly later; meanwhile there are two further points of interest to note about the work. A recently discovered MS. from the collection of the late Arthur Hill has shown that Liszt completed a version of the work for piano and orchestra under the title *Grand Solo de Concert*. This does not contain the central *Andante sostenuto*, and the score, in Liszt's own hand, has all the hallmarks of an inexperienced orchestrator; it is however interesting as showing that as late as 1849 or so Liszt was still fumbling with orchestral technique, whereas only a few years later, as we shall see, he had acquired a very considerable mastery of it. Liszt at any rate seems to have given up the idea of publishing or performing the work as a concerto with orchestra; but in 1856 he made a revised, and in some ways improved version of it for two pianos under the title *Concerto Pathétique,* and in this form it has become fairly well known to-day.

Neither this work nor the Scherzo and March shows Liszt at his best; but the Sonata, like the Faust Symphony of the following year, is certainly one of his masterpieces. The form is that set out above, and the whole of the work is constructed from four fairly short themes which appear in an endless variety of forms. In order to illustrate Liszt's method of " transformation of themes " which was characteristic of him at this period, it will be as well to show this in detail. The first theme is the passage with which the introduction begins:

This is immediately followed by the main theme of the *Allegro energico* (24a)—note the resemblance to Ex. 22 above—and its continuation (24b).

The only other new theme in the first section of the work is this chorale-like passage.

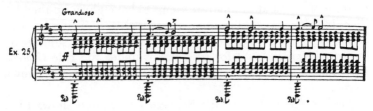

Even the theme which corresponds to the " second subject " in a normal exposition turns out to be a varied form of Ex. 24b.

The whole of the first and third sections of the work are entirely based on these three themes, and they also appear in the central *Andante sostenuto* and in the short coda. In the *Andante* the only additional theme is that at the opening, and part of this too appears in the coda.

Liszt appears to have derived the idea of thematic transformation as a unifying process from Schubert's Wanderer Fantasy, a work of which he was particularly fond, and which he himself transcribed for piano and orchestra in 1851. Schubert's themes in varied forms run through all four movements of the fantasy, which is played without a break, and in addition is based on a symmetrical key scheme—C, E, A flat, C. One can see the strong attraction of this work for Liszt, and many of the works of his Weimar period follow this model, the first piano concerto being a particularly clear example. To Liszt, who wanted to express a series of varying mood pictures, the balancing methods of the classical sonata were of little use; he felt the necessity of creating new forms which would allow him greater flexibility while still maintaining unity. And there is no doubt that in the best works of his middle period, like the Sonata and the Faust Symphony, he triumphantly succeeded in this. The logic of a rigid framework was replaced by the cogency of an emotional argument; but in this case the force of the work will depend on the force of the argument, and therefore Liszt had as many failures as successes in the use of this method. Nevertheless the principle which he established was an important one for future genera-tions; the serial technique of Schoenberg, for instance, uses precisely the methods of Liszt's thematic transformation within the framework of an entirely different language, and it is even possible that future twelve-note composers will turn to forms resembling Liszt's rather than those of the classical composers in the search for a type of framework to correspond to their new methods of expression. In any case Liszt's Sonata remains a landmark in the history of nineteenth-century music, not only as a highly successful application of new technical methods, but as a fine, moving and dramatic work in itself.

Apart from the revised versions of the Transcendental and Paganini Studies, the first two volumes of the *Années de Pèlerinage*, and the Hungarian Rhapsodies, all of which have been dis-cussed in previous chapters, the only other important original piano works of this period are the set of six *Consolations* (1849-50) and the first Mephisto Waltz (1859-60). The latter is more properly considered as a transcription of an orchestral work, and will be returned to later in this chapter; the *Consolations*, which take their title from poems of Sainte-Beuve, are short, charming and original pieces which well deserve performance as a set. The third, in D flat, perhaps the most beautiful of the six, is widely known; the fourth, also in D flat, is based on a theme by the Grand Duchess Maria Pavlovna of Weimar.

The piano transcriptions of this period are also numerous.

They include further works by Beethoven (*An die ferne Geliebte*), Mendelssohn (some of the Midsummer Night's Dream music) and Weber; Hummel's Septet, several songs by Franz, a set of three " Illustrations " from Meyerbeer's *Le Prophète* (one of which is the well-known *Les Patineurs*), some further Berlioz transcriptions, and arrangements of works by younger German composers such as Raff, Nicolai and Draeseke. In addition Liszt transcribed Schumann's *Widmung* in 1848, but subsequently his disagreement with the Clara Schumann–Brahms coterie prevented the publication of further Schumann transcriptions till many years later.

Some of these transcriptions deserve special mention, and have indeed remained in the repertoire owing to their musical qualities. There is a set of six Polish songs by Chopin, which are charming miniatures; there are the nine *Soirées de Vienne*, based on Schubert's waltzes, which are among the most successful things of this kind which Liszt ever did (those who wish to trace the connection between Liszt's pieces and the Schubert originals will find exhaustive information in *The Library of Congress Quarterly Journal*, Vol. VI, No. 2, 1949); and there is the superb piece based on the waltz from Gounod's *Faust*. This opera was produced in 1859, and Liszt's work comes right at the end of the period we are discussing, at a time when he was preparing to leave Weimar in a state of disillusionment after the apparent failure of his hopes. As Mr. Sitwell acutely observes, it has a curious quality of cynical detachment: " Liszt must have seized upon this tune from the most popular opera of the day, determined to make its worldly success his excuse for committing every kind of sacrilege with its body, and yet lifting it, in doing this, on to a higher spiritual plane than it could ever aspire to on its own merits. . . . In this piece he is giving the public their delight and mocking them in that." Mr. Sitwell's description of the work should be read *in toto* as an admirable piece of imaginative writing; it is certainly astonishing that out of Gounod's bouncing and trivial tune Liszt could have produced such a remarkable effect.

Two other groups of transcriptions and fantasies remain to be mentioned, both of continuing importance in Liszt's later life—those based on the operas of Verdi and Wagner.

In 1848 Liszt transcribed the *Salve Maria* from Verdi's *Jerusalem* (better known as *I Lombardi*), and in the following year wrote a fantasy on *Ernani*; this was revised in 1859, for Bülow to play, and published in 1860 together with transcriptions of the *Miserere* from *Il Trovatore* and the quartet from *Rigoletto*—all of them admirably done. Liszt continued to keep track of Verdi, as it were, for the rest of his life, and in later years published transcriptions from *Don Carlos*, *Aida*, and the *Requiem*, as well as a fantasy on *Simone Boccanegra* (1882), his last work in this genre.

Liszt's first Wagner transcription was a brilliant version of the *Tannhäuser* Overture (1848); this was followed by further extracts from *Tannhäuser* and *Lohengrin*, and, later in this period, by the *Spinning Song* from *The Flying Dutchman* and a fantasy on Rienzi. Liszt continued to make Wagner transcriptions (from *Tristan*, *Meistersinger*, *The Ring* and *Parsifal*) up to the end of his life, and in fact did very much the same kind of propaganda work for him that he had done in earlier years for Berlioz.

The question of the musical relations between Liszt and Wagner is a complex one on which volumes could be written. The late Edward Sackville West said, the problem of who influenced whom is like the proverbial question of the hen and the egg. The facts would appear to be approximately as follows:

Liszt first met Wagner in 1841, and first heard *Rienzi* in Dresden in 1844. In 1846 Wagner sent Liszt the full scores of both *Rienzi* and *Tannhäuser*; Liszt had not seen the latter on the stage, nor in fact did he till he produced it himself in Weimar in 1849. At first the relations between the two men were not particularly cordial, but they improved as a result of Wagner's visit to Weimar in 1848. Princess Sayn-Wittgenstein then went to Dresden as Liszt's deputy to see *Tannhäuser*; Liszt made his piano transcription of the overture, and agreed to produce the opera, for the first time outside Dresden (incidentally, at this time he had not seen *The Flying Dutchman* on the stage either). After Wagner's flight from Dresden in 1849, Minna Wagner sent Liszt the score of *Lohengrin* to forward to her husband in Paris; in 1850, at Wagner's request, Liszt gave the first performance of the opera at Weimar.

By 1849, therefore, Liszt had either seen or read the scores of Wagner's first four mature operas; by that time he had himself completed the first versions of the first three symphonic poems, the two piano concertos and the *Totentanz*. The two did not meet again till July 1853, when Liszt visited Wagner at Zürich; after this visit Wagner wrote that he had really got to know Liszt for the first time, and that Liszt had played from MS. some of the symphonic poems (of which about eight had been completed at this time), some of his piano music, and the Faust Symphony. The reference to the last work seems obscure as Liszt did not put it down on paper till the following summer (incidentally Liszt had given Wagner's Faust Overture at Weimar in 1852); but the idea of a Faust Symphony had long been in Liszt's mind, and possibly Liszt played to Wagner his ideas and sketches for the work.* Wagner, who had written no music since *Lohengrin*, now embarked on the composition of *The Ring*, which proceeded as follows: *Rheingold*, November 1853–January 1854; *Walküre*, January–December 1854. Part of 1855 was spent in scoring *Walküre*, and the first two acts were sent to Liszt in October. In the summer of 1856 Wagner underwent a cure at Mornex, and spent the time studying Liszt's symphonic poems—Mazeppa, Orpheus, Les Preludes, Festklänge, Prometheus and Tasso; in September he began the composition of *Siegfried*. Later Wagner wrote to Liszt: "I regard you as the creator of my present position. When I compose and orchestrate I always think only of you. . . . Your last three scores are to consecrate me a musician once more and fit me for the beginning of my second act (*of Siegfried*), which I shall precede by my study of them." This however was broken off in the middle of the second act, and in 1857 Wagner began the composition of *Tristan*. This opera was completed in 1859, and shortly afterwards Wagner wrote the letter to von Bülow in which the celebrated passage occurs: "There are many matters on which we are quite frank among ourselves (for instance, that since my acquaintance with Liszt's compositions my treatment of harmony has become very different from what it was formerly), but it is indiscreet, to say the least, of friend Pohl to babble this secret to the whole world."

I feel that there is no need to treat this question from a partisan point of view. Wagner, a self-centred egotist, found

* Liszt had in fact begun to sketch it in the 1840's.

Liszt useful to him, both artistically and financially, and was no doubt grateful to him in his own way; Liszt genuinely admired Wagner's genius, and certainly learnt a good deal from him. There are of course many parallels between Liszt's and Wagner's music, both at this time and later. We have already seen (p. 52) how a phrase in one of Liszt's songs anticipates the opening of *Tristan*; and there are some curious parallels between *Walküre* and the Faust Symphony, both of which were composed in 1854. The main theme of the Faust Symphony:

appears in this form in the second act of *Walküre*, at the words " Kehrte der Vater nun heim ":

Similarly the " Kiss " motive at the end of *Walküre*:

is strikingly similar to this passage from the finale of the
Faust Symphony:

It would be possible to multiply examples; e.g. the opening
three notes of the *Tristan* prelude are similar to a figure in
the Faust Symphony (Ex. 39 (i)a, p. 79) and there is a
definite resemblance, commented on by both Wagner and Liszt
themselves, between themes in Liszt's *Excelsior!* (1874) and
Wagner's *Parsifal* (1882)—but such reminiscence-hunting is of
little value. What is more important is what each learnt
from the other, and of what advantage it was to each of them.
Liszt learnt from Wagner a stronger sense of form; he was
able to make his music more symphonic and less episodic,
and in addition his command of orchestral writing became
surer and less amateurish—in fact Wagner helped him to
become a really professional composer of large-scale works,
like *Christus* and *St. Elisabeth,* and not merely an inspired
dilettante whose effects might or might not come off. Liszt's
music enriched Wagner's language; it became less conventional,
more pictorial and dramatic, and above all showed him bolder
methods of handling chromatic harmony; one has only to look
at *Tristan* to see how much he learnt in a short time. It was
the 1850s that marked the transition between Wagner's earlier
and later styles, and it was precisely in this period that Liszt
and Wagner were closest together. It was inevitable that they
should drift apart, musically, if not personally, in later years, for
their aims were entirely different; but these years of cross-
fertilisation, as it were, were of the greatest value to both
composers.

There was a further point in which Liszt and Wagner came
together at this period—the question of thematic transformation and the *leit-motiv.* Clearly both composers had been tend-
ing in this direction independently for some years; Liszt, as
we have seen, wrote many early works which are based on
varied treatments of a single theme—possibly the operatic

fantasies, many of which consist of an introduction, variations and coda, all based on the same theme, suggested this method to him. Wagner, in his operas up to *Lohengrin*, did not use *leit-motive* as such; but he did adopt the idea, used consistently by Weber in his later operas, of associating certain music with certain characters and introducing it at suitable dramatic moments—an obvious example is provided by Senta's and the Dutchman's themes in *The Flying Dutchman*. But from *The Ring* onwards his *leit-motiv* technique appears fully developed, and it is surely not too much to suppose that Liszt's methods of thematic transformation provided the impetus for this invention, especially when one considers the dates and times involved.

The question of the development of chromatic harmony is a more complex one. Liszt had already done a great deal of pioneer work in this field, as we have seen, for instance in the extract from *Il Penseroso* (Ex. 16, p. 30), and of course a good many of these " new " chromatic and altered chords had been in use for many years—one can find them in Purcell and Bach, for instance. Even the famous *Tristan* chord is merely a diatonic seventh on the second degree of the scale of E flat minor, taken by itself. But what is new is the treatment; for Wagner approaches and leaves it as if it were in A minor. This question has been fully discussed by Schoenberg in his "Theory of Harmony" (English edition, p. 212), and there is no need for me to repeat what he has said there; but the important step forward was the way these chords were handled, and not the chords themselves. In fact these chords were now treated more or less with complete freedom, and were no longer regarded as discords to be prepared and resolved. A case of particular interest is that of the augmented triad, used either by itself or in conjunction with the whole tone scale. Here again there was no specifically new element; the augmented triad is at least as old as Purcell, and the whole-tone scale had already appeared in Glinka's *Russlan and Ludmila*, where it is used to give a magical dramatic effect. But both Liszt and Wagner made a very considerable use of the augmented triad, and Liszt in particular based several of his later works on a consistent use of the whole-tone scale (see for instance *Der Traurige Mönch*, pp. 96-7). The late Ernest Newman suggested (" A Forgotten Chapter of

History," *Sunday Times*, September 1946) that Liszt was probably
attracted to the use of these elements by the theories of C. F.
Weitzmann, a Berlin musician who published several books,
including one on the augmented triad. This appeared in 1853,
and it is perhaps significant that in the first movement of the
Faust Symphony, written in the following year, not only is the
opening theme based on the augmented triad, but it appears in
one place with the following harmony:

Liszt knew Weitzmann well (he used to play whist with him
in Berlin), but this augmented triad theme had already
appeared in Liszt's sketches for the work dating from the 1840's,
and in the *Grand Chromatic Galop* of 1838, he had already written
this passage:

We are now in a position to discuss Liszt's orchestral works.
As we have seen, Liszt approached the orchestra in an extremely
gingerly manner in his earlier years; but in the later 1840s he
did at last make a serious effort to overcome his lack of knowledge of the technique of orchestration. He began by enlisting
the help of August Conradi (1821-73), a composer of farces and
operettas well known at that time; Liszt wrote out his orchestral
works on three or four staves, together with indications of the
instrumentation required, and from this Conradi prepared a

full score. Liszt then revised this himself, and often the whole process was repeated two or three times until Liszt was satisfied. Conradi was a competent routine composer with little imagination; but from 1849 onwards Liszt enjoyed the help of a far more useful collaborator—Joachim Raff, who came to Weimar expressly for this purpose. The same procedure was followed as with Conradi; but Raff, a composer of far greater imaginative power, was able to make practical suggestions which were of great value to Liszt. However, the final printed versions of all Liszt's orchestral works were revised by Liszt himself, and do represent his own wishes and not those of his collaborators.

The symphonic poem *Tasso* is a good example of this process. Liszt first wrote the music out in short score; from this Conradi made a full score, which was performed on 28 August 1849. Liszt had only completed his own score on 1 August, and there was no time to make corrections based on rehearsal experience until after the performance, when Liszt thoroughly revised the score; this version was used at the second performance on 19 February 1850. But Liszt was still unsatisfied with the work, and now asked Raff to make a new score; this contained many new ideas of Raff's own, but Liszt again made alterations himself, and the new version was tried out at a rehearsal in July 1851. Liszt then put the work away temporarily, but in 1854 he again thoroughly revised it and gave it the form which we know to-day, adding the middle section, which was not in any of the previous versions. By this time he had acquired sufficient mastery of orchestration to be able to work completely independently, and he never used collaborators again (apart possibly from the case of Doppler in connection with the Hungarian Rhapsodies; cf. p. 45). Princess Sayn-Wittgenstein had indeed criticised him sharply in a letter of 25 July 1853 for allowing others to have a hand in the orchestration of his works: "Correction is never as good as original writing," she said; "one can only invent by giving one's thought its first form and first method of expression—a predetermined outline chains the imagination; to a certain extent it sees a road marked out for it, but it discovers no new paths, no new turnings to round out the new forms of thought which one wishes to express." (Those who wish to pursue this object further will find much detailed information in Peter Raabe's "Franz Liszt," Vol. II, pp. 69-79 and 297-310; Raabe also reproduces in facsimile sketches of five different versions of the same passage in *Tasso*.)

We may thus regard the years before 1854 as representing the period in which Liszt was laboriously acquiring a real command of the orchestra, and all the final versions of the symphonic poems date from that year or later. The first twelve symphonic poems were written during this period; they are as follows:

1. Ce qu'on entend sur la montagne.
2. Tasso. Lamento e Trionfo.
3. Les Préludes (d'après Lamartine).
4. Orpheus.
5. Prometheus.
6. Mazeppa.
7. Festklänge.
8, Heroïde Funèbre.
9. Hungaria.
10. Hamlet.
11. Hunnenschlacht.
12. Die Ideale.

The first symphonic poem, also often known as the *Berg-symphonie*, is based on a long romantic poem from Victor Hugo's *Feuilles d'automne*. It seems to have caused Liszt a great deal of trouble, for it was revised four times between 1847 and its final version in 1857. It is a long episodic work, containing many beautiful passages, but rather suffers from being divided into too many self-contained sections. *Tasso* was originally written as an overture to Goethe's play *Tasso*, and the first performance took place on the centenary of Goethe's birthday in 1849. The main theme is a song which Liszt heard a Venetian gondolier sing to the opening lines of Tasso's *Gerusalemme Liberata*; in his introduction he says: " It has been our aim to embody in the music the great antithesis; the genius who is misjudged by his contemporaries and surrounded with a radiant halo by posterity." This idea was a favourite one of Liszt's, and we shall meet it again in *Le Triomphe Funèbre du Tasse* (cf. p. 103), where it clearly has a personal application to Liszt himself; this work is an epilogue to *Tasso*, and is based on some of the same themes, Liszt explains the division of *Tasso* into three parts by saying that they represent " three moments inseparable from his immortal fame. To reproduce them in music, we first conjured up his great shade as it wanders through the lagoons of Venice even to-day; then his countenance appeared to us, lofty and melancholy, as he gazes at the festivities of Ferrara, where he created his masterworks; and finally we followed him to Rome, the Eternal City which crowned him with fame and thus paid him tribute both as martyr and poet." As we have seen, the middle section was not written till long after the rest of the work, and so this explanation is not entirely convincing; it seems more probable that it was inserted for purely musical reasons, to provide a transition between the gloom of the *Lamento* and the brilliance of the final *Trionfo*. Unfortunately, the music of *Tasso* is extremely uneven, and the final section in particular shows Liszt at his most bombastic; but the opening has a fine tragic atmosphere.

Les Préludes, the best known of the symphonic poems, and the only one which has remained in the general repertoire, was originally written in 1848 as a prelude to the choral work *Les Quatre Elémens*. This work, which had been composed a few years earlier, is a setting of four poems by Joseph Autran (1813-77) —La Terre, Les Aquilons, Les Flots, Les Astres—and the themes of *Les Préludes* are taken from it. The choral work, which was orchestrated by Conradi, has never been published, and Liszt later decided to publish the overture as an independent work. He was thus faced with the necessity of finding a " programme " to suit music which had already been written; and he eventually found this in Lamartine's " méditation poétique " Les Préludes. But, in fact, the only thing in common between Lamartine's ode and Liszt's music is that they contain pastoral and warlike elements closely linked together; and therefore Liszt needed to justify the title by writing the well-known preface which appears in the score: " What is our life but a series of preludes to that unknown song of which the first solemn note is sounded by death? . . . " So *Les Préludes* is wrongly described as " d'après Lamartine," and in fact the sequence of moods is Liszt's own. Whether one likes the music or not is a matter of taste; personally I find it an eminently successful work of its kind, though by no means one of Liszt's greatest achievements. Like the other early symphonic poems, it was revised several times, and given its final form about 1854, when it was first performed at Weimar.

Orpheus was also written in 1854, as an introduction to a performance of Gluck's *Orpheus*; Liszt also wrote some closing music on the same themes, which for a long time was thought to be lost, but was later discovered at Weimar. Liszt states in his preface that his work was inspired by an Etruscan vase in the

Louvre representing Orpheus singing to his lyre, and by his music taming the wild beasts and the brutal instincts of mankind; he regarded Orpheus as a symbol of the civilising influence of art. The work throughout has a broadness and nobility which place it high among Liszt's creations, and the final passage for strings echoed by woodwind chords, " gradually rising like the vapour of incense," as Liszt says in his preface, shows a highly daring and original poetry.

Ex. 34

The next symphonic poem, *Prometheus*, was originally written in 1850 as an overture to Liszt's settings of choruses from Herder's *Prometheus Unbound*; the overture and choruses were first performed at Weimar in that year on the occasion of the unveiling of the Herder Memorial there. Five years later both overture and choruses were thoroughly revised and given their present form. Like *Orpheus*, *Prometheus* is a short work and expresses a single idea, that of suffering for the sake of enlightenment, which is symbolised in the legend of Prometheus. Musically the work is unusual for Liszt in that the middle section is in the form of a fugue. The dramatic opening bears some resemblance to that of the *Malediction* concerto (cf. pp. 46-8). Both *Orpheus* and *Prometheus* deserve to be heard far more frequently than they are—which in the case of the latter is practically never.

Mazeppa, as we have seen, is an expanded version of the fourth of the Transcendental Studies. The earlier part follows the *étude* without much more deviation than that necessitated in the transcription of a piano work for orchestra, but the passage depicting Mazeppa's fall is considerably expanded, and a new final section is added, based in part on a " Cossack " theme, which of course symbolises Mazeppa in his new role as Hetman of the Cossacks.*   Here, too, there is a further reference to the struggles of the artist, for Victor Hugo's poem, which is prefixed to the score, describes. him as " Lié vivant sur ta croupe fatale, Génie, ardent coursier," and, as in *Tasso*, triumph only

* Cf. p. 49.

comes after defeat and collapse. *Mazeppa* is unfortunately not one of Liszt's finest creations; though some of the music is exciting enough in a rather obvious kind of way, the march passage has a distinctly vulgar flavour—it is flat and shallow music, as Mr. Sitwell rightly says.

The same must unfortunately be said of *Festklänge*, written in 1853 in anticipation of the celebration of Liszt's marriage with Princess Sayn-Wittgenstein, whose Polish origin is symbolised in the work by a section in polonaise rhythm. Here again we have "effective" music without any great originality of musical thinking. It is more interesting to turn to *Héroïde Funèbre*, for this relates back to the early Revolutionary Symphony (cf. pp. 6-7) and is, in fact, founded on some of the same material. As we have seen, *Héroïde Funèbre* is the only survivor of a planned five-movement symphony; but Liszt now took a rather less ardent view of revolutionary uprisings than he had in his youthful days, and writes in the preface to *Héroïde Funèbre*: "In these successive wars and carnages, sinister sports, whatever be the colours of the flags which rise proudly and boldly against each other, on both sides they float soaked with heroic blood and inexhaustible tears. It is for Art to throw her ennobling veil over the tomb of the brave, to encircle with her golden halo the dead and dying, that they may be the envy of the living." *Héroïde Funèbre* is, in fact, a fine one-movement funeral march of vast proportions, which recalls the shape and feeling of many movements in Mahler's symphonies. The main theme has a distinctly Hungarian flavour:

while in the trio there comes a theme which seems to have originally been part of the Marseillaise in the earlier work:

At the end the music collapses with this remarkable progression:

The work should certainly be heard as a fine example of Liszt's mature style.

Hungaria, like the *Ungaria-Kantate, Funérailles* and some other works of this period, was intended by Liszt as a reply to the poem of homage which the Hungarian national poet Mihály Vörösmarty had dedicated to him after his first Hungarian concert tour in 1840; and as we have seen, it is partly based on the Heroic March in the Hungarian style written in that year. Unfortunately, in spite of many beautiful and exciting episodes, it suffers from the over-elaboration and inordinate length of many of the works of the Weimar years; it is a pity that Liszt never revised and shortened it, for it contains much good material—as it stands, the Heroic March is far the better piece of the two. No programme is attached to *Hungaria*, which may best be regarded as a Hungarian Rhapsody on an extended scale.

The next symphonic poem, *Hamlet*, however, is one of Liszt's masterpieces. It was written in 1858 as an overture to Shakespeare's play; Liszt had been greatly impressed two years earlier by a performance of the play in Weimar with Bogumil Dawison in the title role. He felt that Dawison had presented a concept of Hamlet which was entirely new to him—not the indecisive dreamer, as described, for instance, by Goethe in "Wilhelm Meister," but as a gifted enterprising prince who was only awaiting the right moment to carry out his revenge on his uncle and fulfil his political ambitions—and he felt that this interpretation was far more in line with Shakespeare's construction of the play as a whole. In this interpretation the part of Ophelia becomes a very secondary one; she is not the usual romantic heroine dear to nineteenth-century conceptions. "Yes, she is loved by Hamlet," Liszt wrote, "but Hamlet, like every exceptional person, imperiously demands the wine of life and will not content himself with the buttermilk. He wishes to be understood by her without the obligation to explain himself to her. ... She collapses under her mission, because she is incapable of loving him in the way he must be loved, and her madness is only the decrescendo of a feeling, whose lack of sureness has not allowed her to remain on the level of Hamlet."

Liszt's symphonic poem is a psychological study of Hamlet's

character, without any particularly programmatic elements. It consists of a slow introduction, a dramatic and violent Allegro, and a slow final section, partly referring to the introduction, and ending in the manner of a funeral march. In the middle of the Allegro there are two short interludes (which as a matter of fact were an afterthought on Liszt's part) directed " to be kept as quiet as possible and sound like a shadowy picture, referring to Ophelia." It is a short and concise work (it plays about ten minutes); in it nothing is wasted, every point is made with clarity and precision, and a remarkable psychological portrait emerges. Of all Liszt's symphonic poems this is the one which most merits revival and frequent performance; yet even in Liszt's lifetime it was not properly appreciated, and the first performance did not take place till 1876 in Sondershausen.

The idea of *Hunnenschlacht* came to Liszt on seeing a reproduction of the enormous fresco by Wilhelm von Kaulbach representing the battle in 451 A.D. in the Catalaunian fields between the Huns under Attila and the Christian emperor Theodoric's forces for the possession of Rome. The battle, which was finally won by the Christians, was so fierce that in the dusk the exhausted survivors seemed to see the spirits of the slain continuing the combat in the sky. Liszt at this time even considered writing a series of symphonic poems after Kaulbach's frescoes to be called " The History of the World in Sound and Picture," and to include the Tower of Babylon, Nimrod, Jerusalem, and the Glory of Greece. But he apparently failed to find a writer who could translate the content of Kaulbach's pictures into a poetical form which would give him the basis for his music—a curious need, but one which Liszt evidently felt, for in *Hunnenschlacht* he wrote out his own programme in words, to act as a kind of transitional stage between the picture and the music.

The first half of *Hunnenschlacht* is a battle scene painted with great power and originality, beginning as it were in darkness, and gradually increasing in fury to a tremendous climax; Liszt originally intended to return from this to the mood of the opening, and to end with the picture of the spirits battling in the air. But he later decided on a less directly pictorial treatment; the battle in fact ends halfway through the work, and thereafter the Christian chorale *Crux fidelis*, beginning quietly on the organ, gradually leads up to a triumphant ending—here the universal victory of love over hate in men's hearts is symbolised. rather than any one physical battle. Musically this makes the construction of the work rather episodic, but the representation of the battle is one of the finest things Liszt ever wrote.

The last of these first twelve symphonic poems, *Die Ideale*, is based on a poem by Schiller, and extracts from this poem are printed in the score. The work was originally designed as a symphony in three movements, but was later compressed into a three-movements-in-one form akin to that of the Sonata—Andante introduction, energetic first section, slow dolorous second section, scherzo-like third section and a short coda or finale, " Maestoso con somma passione." In fact the musical construction became more important than the poetical aspect; Liszt changed the order of appearance of his extracts from Schiller's poem, and towards the end of the work wrote: " I have allowed myself to add to Schiller's poem by repeating the motives of the first movement joyously and assertively as a apotheosis." Like " Ce qu'on entend " the work is long and episodic, but contains many fine ideas; this theme seems to have served as an inspiration to many later composers.

Ex. 38

This review of Liszt's first twelve symphonic poems will have shown, I hope, that though many of them contain pictorial and programmatic elements, Liszt's approach was fundamentally different from that of Saint-Saëns, Strauss, Dvořák, Sibelius and others who have used this form. In the symphonic poems Liszt wished to expound philosophical and humanistic ideas which were of the greatest importance to him, and many of which were connected with his personal problems as an artist (*Tasso, Orpheus, Prometheus*). Thus where Beethoven's symphonies may be said to be concerned with undefined philosophical problems which cannot be expressed accurately in words, Liszt in this new form was trying to represent more explicit problems which had been set out in many cases by writers or painters; he was not interested in the minute pictorialism into which the symphonic poem later degenerated, nor, in the first place, in " telling a story " in music; the story, if any, to him was merely the symbol of an idea. This is not an easy conception to fulfil; and as we have seen, Liszt scored as many failures as successes; but the invention of the symphonic poem was a landmark in musical history, and Liszt rightly felt that with it the scope and expressive power of music had been greatly widened, if not always deepened.

Apart from a number of marches and other occasional pieces of no great musical significance, Liszt's remaining purely orchestral works of the Weimar period consist of the Faust and Dante Symphonies and the *Two Episodes from Lenau's Faust* (*Les Morts* will be discussed in the next chapter, in connection with the other *Odes Funèbres*). Liszt had been introduced to the first part of Goethe's Faust in the early 1830s by Berlioz (in his memoirs Berlioz relates that this took place on the occasion of their first meeting, the day before the first performance of the *Symphonie Fantastique* in December 1830). The subject clearly appealed to Liszt from the first, and he was given a further stimulus by the production of Berlioz' *Damnation of Faust* in 1846 —the Faust Symphony is in fact dedicated to Berlioz. But at the same time he had some misgivings—shortly before starting work on the Symphony he wrote to the Princess: " Anything to do with Goethe is dangerous for me to handle "—and in a letter written fifteen years later he explained this to some extent. " In my youth," he said, " Faust seemed to me a decidedly bourgeois character. For that reason he becomes more varied, more complete, richer, more evocative [than Manfred, on whom Liszt also seems to have wanted to write a musical work at one time]. . . . Faust's personality scatters and dissipates itself, he takes no action, lets himself be driven, hesitates, experiments, loses his way, considers, bargains, and is only interested in his own little happiness." Clearly, then, it took him some time to absorb and appreciate the real character and significance of Faust; but once he did start he found the way clear before him, and in fact the whole enormous work was written in the astonishingly short space of two months (August–October 1854).* Liszt was then at the height of his powers as a composer; he had solved his formal problems in the Sonata of the previous year, and in addition had at last achieved mastery of orchestral technique—the whole of the Gretchen movement for instance, with its delicate and original orchestration, was written straight out in full score and hardly touched since. Above all, the work must have been very near to his heart; the first movement, Faust, with its many changes of mood, has all the appearance of being a self-portrait; the second movement expresses all his love and admiration for women; and the finale is a supreme manifestation of that diabolism which is to be found in all periods of his life from the first to the last.

Formally the work is constructed on much the same lines as the Sonata, in that the first and third movements are inter-related, while the coda of the third movement is based on the themes of the second; but the movements are conceived as separate entities, and the whole structure is of course far larger and more complex. (A more detailed analysis of both the Faust and Dante Symphonies will be found in " The Symphony," ed. Robert Simpson, Pelican Books, 1965). The first movement consists of a long exposition, a brief development section (a good deal of " development " has already gone on in the exposition), a very much shortened recapitulation, and a short coda. It is based on five main themes, all symbolising different aspects of Faust's character. The first, with its " magical " whole-tone flavour, has already been quoted (Ex. 28, p. 65, and Ex. 32, p. 68); the second, which immediately follows it in the slow introduction, bears an obvious resemblance to the themes already quoted from the *Grosses Konzertsolo* (Ex. 22, p. 57) and the Sonata (Ex. 24, p. 59). This theme, with its ability to change its character radically at will, constitutes one of the most important elements in the work; Ex. 39 gives some of its varied manifestations.

* It had, however, been partly sketched in the 1840's (cf. p. 64), and it was considerably revised between 1854 and 1861.

The remaining Faust themes consist of a stormy *Allegro agitato* passage for the strings, a flowing emotional theme based on a descending scale which immediately follows it on oboe and clarinet, and a *Grandioso*, martial theme which appears later on

Ex. 39

the brass. The whole of the first movement is based on these themes, and in addition all of them except the first appear in the slow movement, Gretchen, with the addition of two new themes which symbolise Gretchen herself. This movement is in the usual ABA form, the main sections being mostly based on the Gretchen themes, while the Faust themes chiefly appear in the middle section (representing the lovers coming together) and the coda. This movement contains the one definitely pictorial passage in the work—a little interlude portraying the scene where Gretchen plucks the petals of a flower, murmuring to herself " He loves me—he loves me not—he loves me! " Otherwise the symphony is what its title describes—three character studies after Goethe.

The finale, Mephistopheles, is in some ways the most remarkable movement. It is based entirely on the Faust themes, but in a parodied form—Mephistopheles, the spirit of negation, can only destroy, not create. There is one new theme—the " pride " motive from the *Malediction* concerto (Ex. 19, p. 48), and Gretchen's theme makes a brief appearance in the middle of the movement in its original form—she alone cannot be touched by the Devil. The whole atmosphere is brilliantly conceived and portrayed; even to-day it sounds sinister enough. Liszt originally intended to end the work purely orchestrally, with the disappearance of Mephistopheles and a reference to the Gretchen theme and the "heroic" Faust motive; but three years later he added a coda in the form of a choral setting of the Chorus Mysticus which closes Part II of Goethe's Faust, and there is no doubt that he was right to do so, for the entire work is summed up and given its true perspective by this solemn and triumphal music. Liszt was truly inspired when he wrote the Faust Symphony; it expresses every variety of mood with the utmost clarity and dramatic emphasis, yet one never feels that the music is forced or artificial. It simply poured out of him quite naturally; though the symphony lasts over an hour one does not get the impression that it is overlong for what it has to say, for it is all deeply and genuinely felt. Many think, and I would agree, that in this work Liszt produced his masterpiece.

The Dante Symphony is more open to discussion, though curiously enough Liszt here felt a stronger instinctive affinity with his subject. He had read Dante with Marie d'Agoult in the 1830s, and the first version of his Dante Sonata dates from as early as 1837; by 1847 he had already sketched out the principal themes of the symphony, and intended at that time to get the painter Buonaventura Genelli to design lantern slides to be shown during the performance of the music. But he did not start on the Symphony in earnest till the summer of 1855; and although it is a much shorter work than the Faust Symphony, it took him much longer to complete—about a year, in fact. His original idea had been to write three movements. Inferno, Purgatorio and Paradiso, to correspond with the three parts of Dante's *Divina Commedia*; but unfortunately Wagner managed to persuade him that no human being could dare to portray the joys of Paradise, and therefore Liszt wrote instead a Magnificat which replaces the third movement. The result is a somewhat unbalanced work; and the listener is left in the transitional state of Dante at the end of the Purgatorio section, where he gazes up at the heights of Heaven and hears its music from afar.

Nevertheless the Dante Symphony contains a good deal of fine music, particularly in the first movement (on which, incidentally, Tchaikovsky's *Francesca da Rimini* appears to have been closely modelled). It begins with a musical setting in the trombones of the inscription written over the gates of Hell:

> Per me si va nella città dolente;
> Per me si va nell'eterno dolore;
> Per me si va tra la perduta gente. ....
> Lasciate ogni speranza voi ch'entrate.

The last line is actually made into a musical theme, which plays an important part in the movement.

Then a descending chromatic theme begins the first section of the movement, the representation of Dante's " Strange tongues, horrible cries, words of pain, tones of anger " of the whirlwind of Hell. The movement is in ABA form; like Tchaikovsky after him, Liszt gives in the middle section a portrayal of the unhappy Paolo and Francesca, and one of their themes is again based on a verbal quotation:

> Nessun maggior dolore
> Che ricordarsi del tempo felice
> Nella miseria.

(Incidentally the construction of this work is rather more orthodox than that of the Faust Symphony, and there is little use of " transformation of themes "; but many of the motives, like Ex. 41 above, are based on a descending scale—similar themes may be found in the second movement as well).

After the Paolo and Francesca episode, the storm returns,

and eventually brings the first movement to a violent conclusion. The second movement begins with an introduction which represents Dante coming up from Hell into the light of the stars and seeing the dawn rising like " the sapphire of the orient." This beautiful passage leads to the main part of the movement, which represents the souls in Purgatory undergoing the trials which will fit them for Heaven. The central section is a fugue, again on a descending scale theme, which rises to a passionate climax; finally the women's voices enter quietly with the Magnificat. The ending is tranquil and joyful, with long-held ethereal chords; a second, fortissimo ending, composed at the Princess' instigation, is definitely inferior and destroys the whole atmosphere of Dante's original passage.

The Two Episodes from Lenau's *Faust* are the last important orchestral works of the Weimar years, having been written about 1860. Lenau's work is a long dramatic poem which contains many episodes omitted from Goethe's version of the Faust legend. The first of the two interpreted by Liszt is *Der nächtliche Zug*—The ride by night. It is a warm spring night, dark and gloomy, but the nightingales are singing. Faust enters on horseback, letting his horse quietly saunter on. Soon lights are seen through the trees, and a religious procession approaches, singing the chorale *Pangue lingua*. The music rises to a climax, the procession passes on, and Faust, left alone, weeps bitterly into his horse's mane. This fine descriptive piece, with its wonderful evocation of atmosphere, for some reason is hardly ever performed, though there is nothing exaggerated or overdone about it; in fact it shows what complete mastery of orchestral writing Liszt had by now attained. Its companion piece, however, The Dance in the Village Inn, better known as *the* Mephisto Waltz (there are, in fact, three others, which will be discussed in the next chapter), is frequently played, both for orchestra and as a piano piece—both versions were composed at approximately the same time. Faust and Mephistopheles enter the inn in search of pleasure; the peasants are dancing, and Mephisto seizes the violin and intoxicates the audience with his playing. They abandon themselves to love-making, and two by two slip out into the starlit night, Faust with one of the girls; then the singing of the nightingale is heard through the open doors. There are two alternative endings which come at this point; in the one usually played the music of the dance returns, and works up to a brief and violent climax; but the second ending is dramatically more effective. There is a sudden fortissimo in the whole orchestra, and then the music dies away in muttered tremolos; the quotation in the score reads: " They sink in the ocean of their own lust." The first Mephisto Waltz is a brilliant dramatic piece which rightly deserve its place in the repertoire; the middle section, with its chromatic theme, looks forward to the harmonic methods of Scriabin:

Liszt's works for piano and orchestra completed during this period consist of the two concertos, the *Totentanz*, the Fantasy on Beethoven's *Ruins of Athens*, the Hungarian Fantasia, and the transcriptions of Schubert's Wanderer Fantasia and Weber's *Polonaise Brillante*, Op. 72. The two concertos date in origin from the 1830s, the main themes of the first concerto being found in a sketch book dated 1830 itself; Liszt appears to have worked on it again during the late 1840s, and further alterations were made in 1853 and 1856. The first performance took place on 17 February 1855 in Weimar; Liszt was the soloist and Berlioz the conductor. Perhaps because of its origin in Liszt's early youth, the first concerto is not an entirely successful work; the themes lack real distinction, in spite of their brilliant and ingenious treatment. There is too much of the glitter of the virtuoso about this work—though admittedly this is accentuated by those modern pianists who emphasise its brilliance at the expense of its dignity and play it at about twice the speed intended. (The recordings of both concertos, made shortly before the second world war by Sauer as pianist and Wein-gartner as conductor—both of them Liszt pupils—are salutary in this respect.) The four short movements are played without

a break—on the model of the Wanderer Fantasy—and the
theme of the lyrical slow movement:

Ex. 43

is transformed into that of the final march:

Ex. 44

In fact the whole work is integrated thematically in the way we
have observed in the Sonata and the Faust Symphony. Another
interesting point is Liszt's use of solo orchestral instruments in
conjunction with the piano—for instance the long passage with
clarinet solo in the first movement:

Ex. 45

This is typical of Liszt's approach to the orchestra in general; his aim was always clarity, and he did not care for the unrelentingly opaque texture of Wagner and other later nineteenth-century composers. In fact, he was a protagonist of " chamber music for full orchestra "—thus anticipating the methods of modern composers like Schoenberg and Stravinsky—and he reserved his heavy scoring for selected *tutti* passages. It is true that he did overscore in some of these, and that his brass writing is sometimes heavy and tasteless, but his aim was always to get his thought over to the audience in as simple a way as possible, and he did not mind if the sonority was often thin and bare—he was much more interested in obtaining variety and colour than in a " safe " but monotonous fullness of texture. One has only to compare his scores with those of Schumann or even Brahms to see the difference of approach at once.

The second concerto is a very much more successful work than the first. It was first written in 1839, and revised at various times between then and 1861; the first performance, also at Weimar, took place in 1857. Here the themes are far more interesting and capable of development; for instance the poetical opening:

a strange theme for the beginning of a piano concerto, one might think, for clearly on the piano it would sound nothing—but Liszt never gives it to the piano in this form; it is always split up into arpeggios or broken chords, or decorated in some way, and it is in touches like this that Liszt shows himself a master. The transition, too, from the slow introduction to the Allegro which follows is also supremely well done, so that one arrives in the quick tempo before one knows where one is, so to speak. The one really weak passage in the work is the transformation of Ex. 46 into a march in the finale; it has all the pompous vulgarity of second-rate military band music, and is quite out of place in this beautiful concerto, which certainly stands high among Liszt's achievements.

But the most remarkable of all these works is the *Totentanz*. This was inspired as far back as 1838, when Liszt saw Orcagna's frescoes " The Triumph of Death " in the Campo Santo at Pisa. The first version (which Busoni edited and published in 1919) was completed in 1849; the work was again revised during the next ten years, and was first performed and published in 1865. It consists of a series of variations on the *Dies Irae*, and is a work of astonishing dramatic power. This passage from the opening cadenza shows how far Liszt was prepared to go in the simultaneous use of apoggiaturas and their resolutions.

Though by no means diabolistic throughout—there are some fine lyrical passages—it shows Liszt in the same mood as the finale of the Faust Symphony, as a master of dramatic expression.

The remaining works in this category are not of great importance. The *Ruins of Athens* Fantasy is little more than a brilliant showpiece—which also applies to the transcription of the Weber Polonaise—and the Hungarian Fantasia is an effective arrangement of the 14th Hungarian Rhapsody. The arrangement of Schubert's Wanderer Fantasia is of more interest; it is certainly not what Schubert might have written, yet in a curious way it keeps faithful to the spirit of Schubert. Particularly effective, for instance, is the opening of the last movement, where the theme is announced fugally, beginning fortissimo in bass octaves. A lesser arranger might well have given this to the lower strings and brought in the higher instruments successively with each subsequent entry; Liszt rightly gave the entire fugal exposition to the solo piano, knowing that it alone could give the effect that Schubert intended, and reserved the orchestra for the subsequent passage. The soloist is in fact the dominant partner throughout, and the orchestra either supports or provides a contrast to his part. The whole question of an arrangement of this kind

depends on whether the letter or the spirit is of greater importance; but Liszt, always a sincere admirer of Schubert, has
certainly been faithful to his original " in his fashion."

Another category of works on which Liszt embarked for the
first time during these years were compositions for organ. He
was to write very many more of these in the post-Weimar
years, when he was becoming increasingly interested in church
music; but during this period he completed two works which
are in fact his greatest compositions for the medium—the
Fantasy and Fugue on " Ad Nos, ad Salutarem Undam," and
the Prelude and Fugue on the name of B.A.C.H. The " Ad
Nos " Fantasy is in fact No. 4 of the " Illustrations " from
Meyerbeer's *Le Prophète* (cf. p. 62), but it is certainly not an
operatic fantasia. It is based on the chorale sung by the three
Anabaptists in the first act of the opera, where they call the
people to seek re-baptism in the healing water. It is a big work,
lasting over half an hour in performance; it falls into three
parts, which are played without a break. The chorale theme is
heard in part at the beginning of the opening section, which is
a true Fantasia with many changes of mood; then follows a
beautiful and poetical Adagio movement, which begins with a
quiet statement of the chorale in full. At the end of the Adagio a
brilliant rushing passage introduces the fugue, which, following
Liszt's usual practice, does not remain strictly fugal for long;
and the final climax contains a triumphant statement of the
chorale theme. (Incidentally this theme is Meyerbeer's own,
and not, as has frequently been stated, a traditional chorale.)
The work is certainly an admirable addition to organ literature,
and, heard in the right circumstances, can sound overwhelming;
there is also an excellent piano transcription of it by Busoni.

The BACH Prelude and Fugue was originally written in 1855
and revised in 1870; this later version was also transcribed for
piano by Liszt. It makes an advanced use of the chromaticism
inherent in the motto theme; the opening of the fugue (Ex. 48,
p. 88) in particular contains a long passage which is shifting in
tonality so frequently that it is impossible to say what key it is
really in. This type of chromaticism, based to some extent on the
chord of the diminished seventh, was considerably influenced by
Bach's own use of chromatic harmonies, particularly in some
of the chorale harmonisations and chorale preludes. Liszt was

much attracted to it in several works of a slightly later period, such as the " Weinen, Klagen " prelude and variations and some of the later organ and choral works. Liszt, of course, used these methods in a much more radical manner than Bach, who

Ex. 48

always preserved his sense of tonality; but with Liszt we can see the beginning of that sliding chromaticism which eventually weakened the tonal system at the end of the century, in the works of composers like Reger, to such an extent that tonal analysis hardly became possible any more; and this in turn paved the way for the atonal music of Schoenberg and his followers. (As early as the 1830s Liszt had worked on the idea of a possible *ordre omnitonique* which might be destined to supersede normal tonality; the MS. of a *Prélude Omnitonique* by him was shown at a London exhibition in 1904, but it has unfortunately not so far been possible to trace its whereabouts.) So this Prelude and Fugue may be regarded as a more or less direct link between Bach and Schoenberg; and apart from its experimental side, it is in any case an extremely interesting and effective piece.

The story of Liszt's abortive attempts to complete a mature opera also dates from this period. In view of his lifelong fascination by the opera it was inevitable that he should at some time attempt to write one himself, and the surprising thing in a way is that he never finished such a work (apart, of course, from the youthful *Don Sanche*; cf. p. 3). But his correspondence shows that he was full of ideas for operas, and some of them even got within some distance of fruition. Both his symphonies were originally thought of as operas; in 1845 he planned a stage version of Dante's *Divina Commedia*, with a text by Autran (cf. on *Les Préludes*, p. 71), and in 1850 he had ideas for a Faust opera in collaboration with Dumas. He even returned to the latter idea, this time in collaboration with Gérard de Nerval, as late as 1854, the year of the composition of the Faust Symphony! However he did get further than ideas with other operatic projects; in 1846 he wrote to his Weimar patron, Grand Duke Carl Alexander: " The aim which is important to me before and above everything at this moment is to conquer the theatre for my creations, as I have conquered it during the last six years for my personality as an artist; and I hope that next year will not pass without my arriving at a more or less decisive result in this new career. You would not be able to believe, my Lord, the time and patience I shall need to carry through my libretti to perfection (for I am impertinent enough to be working at two Italian operas at once, and to announce my intention of giving birth to twins at the first attempt!). But at last, God be praised, here I am occupied with this task, and apart from unforeseen difficulties, from next May onwards the Italian company of the Kärntnertor theatre will be bursting their lungs with ' felicità! dolore! valore ed amore! ' on cantilenas in my manner! "

" Unforeseen difficulties " did arise, and the only tangible result of all this activity seems to be 111 pages of sketches in the Weimar Liszt Museum for an Italian opera, *Sardanapale*, to a text adapted from Byron, probably by Rotondi. The plot is as follows:

Act 1. The priest Beleses arouses the Assyrian people against their king Sardanapalus, who has forgotten his kingdom, his wife and his children for the love of the slave girl Myrrha. The rebels wish to replace Sardanapalus by the satrap Arbaces, who also loves Myrrha; he declares his love to her, and wishes to win her for the cause of the conspirators, but is rejected by her. Sardanapalus enters and has Arbaces arrested, but pardons him at the request of Myrrha, whose nobility wins the people's homage. In the midst of the jubilation Queen Zarina enters; she demands that Sardanapalus shall give up Myrrha—otherwise she herself will leave the kingdom. The act ends with an ensemble, in which Beleses promises to avenge the queen.

Act 2, Scene 1. Zarina's room. Beleses and Arbaces offer her their help. The queen gives them a ring which permits them entry to the palace. They swear to her, hypocritically, to kill Myrrha but to spare Sardanapalus; in reality Beleses intends to overthrow Sardanapalus, and Arbaces to win Myrrha for himself. In the second scene Sardanapalus and Myrrha are feasting in a brilliantly lighted room; at the height of the revels the torches suddenly go out, and the ghosts of three former Assyrian rulers, Belus, Nimrod and Semiramis, appear and warn Sardanapalus of his approaching downfall. However, he takes no notice of the warning and orders his companions to continue with the feast. Then the uproar of the conspirators is heard outside, and a wounded soldier brings news that the city has been captured. Sardanapalus seizes his weapons and leaves for the battle with Myrrha and his followers.

Act 3. Sardanapalus has been defeated in the battle, but his troops still hold the palace. He is reconciled to Zarina, and sends her, together with Myrrha and his children, out of the country in a galley for safety. He himself waits in the besieged palace; but when he hears that the last of his hitherto faithful satraps has gone over to the enemy, he resolves to die. Myrrha suddenly enters; she has fled from the galley in order to be with him in his hour of greatest danger. Together they mount a pyre loaded with costly spices, kindle it and meet their death.

The plot is certainly in the old sensational operatic manner, and Liszt's sketches for the music also reflect the style of the Italian opera of the time; but while they lean towards Bellini, Meyerbeer or the Wagner of *Rienzi* rather than *The Flying Dutchman* or *Tannhäuser*, the music is on the whole simpler and more symphonically constructed than its Italian models. The sketches consist merely of the vocal line with a piano accompaniment containing some indications of the orchestration, and are for the most part extremely fragmentary; Ex. 49 is a typical example of the style.

Although *Sardanapale* was never completed, Liszt still did not

despair of writing an opera, and the Liszt Museum contains
several libretti which he appears to have considered at one time
or another, including two different versions of scenes from
Byron's *Manfred* (one of them prepared by Cornelius); *St.
Hubert*, a romantic opera in three acts; *Semele*, a one-act
opera after Schiller; and " *Kahma—La Bohémienne* " (a pity that
this was never written!) with a book by Otto Roquette, who
also wrote the text for *St. Elisabeth*. In addition, between 1856-8
he was thinking seriously of writing a Hungarian opera.
János or *Janko*, with a text by Salomon Mosenthal after a poem
by the Hungarian writer Carl Beck, and even wrote that the
first performance would take place in Budapest in 1859 or
1860; but in the end he made objections to the libretto, which
was eventually set by Rubinstein under the title *The Children
of the Heath*. At the same time, at the suggestion of Princess
Sayn-Wittgenstein, he was trying to persuade the writer
Friedrich Halm (the pen-name of Baron Münch-Belling-
hausen, later Intendant of the Vienna Court Theatre) to make
a libretto on the subject of Joan of Arc; but Halm was
frightened of the precedent of Schiller, and Liszt could find no
other satisfactory librettist. (He had, however, already set
Dumas' " *Romance dramatique* " *Jeanne d'Arc au Bûcher* as a
song with piano, and about this time he orchestrated it, in a
form which almost amounts to an operatic aria.)

So the only mature work of Liszt which has appeared on the stage is the oratorio *St. Elisabeth*; but here the stage version was undertaken against his wishes, and he in fact refused to be present at the first stage performance at Weimar in 1881—rightly so, for the dramatic tempo of the music is far too slow for stage action. We must regret that Liszt never completed a mature opera—perhaps the overpowering dramatic genius of his friend Wagner made him lack confidence in this direction—but it is at any rate interesting to speculate on what might have been.

The remaining works of this period consist of choral works, sacred and secular, songs and recitations. The sacred choral works are few in number; the first one of importance is the Mass for four-part male chorus and organ, written in 1848. Liszt in later years regarded this work as a step on the way to higher things; it is simple and liturgical in style, and Liszt wished it above all to express " religious absorption, Catholic devotion and exaltation. The church composer is also preacher and priest," he wrote, " and where words cannot suffice to convey the feeling, music gives them wings and transfigures them." This was one side of his attitude to church music, and it may be found in a number of his simpler and smaller works of this kind. He always wished to be regarded as a loyal son of the Catholic church, and therefore in many works deliberately restricted himself to a style which would be in keeping with church traditions; he had made a study of Gregorian music, and often incorporated elements of it into his church works—for instance, the Gregorian intonation was used consistently by him as a

symbol of the Cross in many works, including this Mass, the Gran Mass, *St. Elisabeth*, the final chorus of the Dante Symphony and the *Hunnenschlacht*—and in the later *Via Crucis* (cf. p. 119) it occurs throughout the work as a kind of *leitmotiv*. Further Gregorian elements may be found in the revised version of this male chorus Mass, made in 1869.

The other side of his approach to church music may be seen in his setting of Psalm 13 (1855) for tenor solo, chorus and orchestra. Here the dramatic technique of the symphonic poems is applied to a religious subject, and the result is a fine, intensely exciting and moving work. It has more in common with later dramatic settings of religious texts, like Verdi's Requiem, or even a modern work like Kodály's *Psalmus Hungaricus*, than with the usual oratorios and cantatas of the time, and hence Liszt was accused of being an " effectmonger." Nothing could be more unjust; Liszt felt the suffering and the joy expressed in the words with great directness, and he simply wished to give it the fullest expression possible. And the fact that it has still lost none of its power to-day shows that he was well justified.

Liszt wrote several other psalm settings, of which three date from this time—those of Psalms 18, 23 and 137. They come at the very end of this period (1859-60), when Liszt's thoughts were turning more and more in the direction of church music. The 18th Psalm is set for male chorus and orchestra, with alternative versions for organ and for wind orchestra; Liszt himself described it as " very simple and massive—like a monolith." The other two psalms are for smaller combinations; Psalm 23 is for solo voice, harp and organ, with an alternative version including a male chorus, and Psalm 137 (" By the waters of Babylon ") is for soprano solo, women's chorus, solo violin, harp and organ. The latter appears to have been inspired by Eduard Bendemann's painting, " The mourning Jews by the waters of Babylon," and contains many distinctly pictorial effects. It is certainly one of the more interesting of Liszt's religious works.

During this period Liszt also received the initial inspiration for his two greatest oratorios, *St. Elisabeth* and *Christus*, but as they were not completed until the Rome years it will be best to discuss them in the next chapter. However it may be noted that Liszt began the composition of *St. Elisabeth* in 1857, and also wrote in these years the Beatitudes and the Pater Noster which now form part of *Christus*. His only other important religious work of this period is the Gran Mass, written in 1855 for the dedication of the Basilica at Gran (Esztergom) in Hungary. This work caused considerable discussion at the time; Liszt was accused of imitating Wagnerian methods in it, and even of trying to " smuggle the Venusberg into church music." The

Mass certainly gives no cause for alarm in these days; it is highly dramatic and full of feeling, but so is a good deal of church music written earlier (including, for instance, Beethoven's Missa Solemnis) and since. Hanslick wrote: " We do not raise the slightest doubts about the religious feelings of the composer," but went on to imply that Liszt was like a clever actor trying to find new points to bring out in an important rôle which he had played too frequently—i.e. that he was merely trying to do something different from what anyone else had done before. This is unfair to Liszt, who certainly felt very strongly what he was writing, even if his feelings were sometimes superficial. To us to-day the Mass makes a sincere, if not an overwhelming impression; it contains many beautiful moments, particularly in the opening Kyrie, and there are exciting passages in the Gloria and Credo. This passage from the Kyrie deserves quotation:

Ex. 51

Three of the secular choral works of this period are of importance. The first in date is the *Ungaria-Kantate* of 1848; as we have seen (p. 74) this was intended as a patriotic tribute, and Liszt seems also to have had some idea of including it in his projected Revolutionary Symphony at this period (cf. pp. 6-7). Liszt wrote the work for baritone solo, mixed chorus and piano, and from this Conradi prepared an orchestral version, the unpublished MS. of which is in the Liszt Museum; the work was first performed at Weimar in 1912, but not published till 1961 in Budapest. The choruses, originating from Herder's *Prometheus Unbound* (cf. p. 72), were first performed as part of a stage representation, and later Pohl wrote a linking text for concert use—this version was first performed in 1857. The work is written for six soloists, mixed chorus and full orchestra, and lasts nearly an hour; Liszt also arranged the Pastorale (Schnitter-Chor) for piano, and in this form it has become moderately well known. Liszt's setting for male voices of *An die Künstler* was originally written in 1853, and the accompaniment was scored for wind instruments by Raff; later in the year Liszt re-arranged it and scored it for full orchestra, being at any rate partly influenced by the unfavourable reception which the original version received at its first performance. The final version was first performed in 1857, and was a great success. The text consists of some lines from Schiller's poem " Die Künstler "; in 1854 Liszt sent the work to Wagner, who wrote to him as follows: " You had in me a somewhat adverse judge of this composition—I mean, I was not in the mood for it. . . . It is more or less a didactic poem. In it there speaks to us a philosopher who has finally returned to art, and does so with the greatest possible emphasis of resolution. . . . I could not at any price write a melody to Schiller's verses, which are entirely intended for reading. These verses must be treated musically in a certain arbitrary manner, and that arbitrary manner, as it does not bring about a real flow of melody, leads us to harmonic excesses and violent efforts to produce artificial wavelets in the unmelodic fountain." Nevertheless, on looking at the work as a whole, he goes on: " This your address to the artists is a grand, beautiful, splendid trait of your own artistic life. I was deeply moved by the force of your intention. . . . You have done well in drawing Schiller's lines out of their literary existence and in proclaiming them loudly and clearly to the world with trumpet sound. . . . I at least know nobody who could do something of this kind with such force." Wagner was not much given to praising other composers' works, and this spontaneous tribute is at least impressive. Incidentally, Liszt made use of themes from *An die Künstler* in the symphonic poem *Die Ideale* (cf. p. 76) and also in the *Künstlerfestzug zur Schillerfeier* of 1859.

About fifteen of Liszt's songs date from this period, during which he also revised and improved the majority of his early songs, and in 1860 a fairly comprehensive collection of them, *Gesammelte Lieder*, was published. The new songs are all settings of German poems; among the most successful are two Heine settings, " Ein Fichtenbaum steht einsam " and " Anfangs wollt' ich fast verzagen," with their sensitive feeling for atmosphere; the well - known " Es muss ein Wunderbares sein "; " Die drei Zigeuner " (Lenau), a remarkable study in characterisation with a strong tzigane flavour; and the lyrical and passionate " Jugendglück." There is no radical change from the style of the earlier songs; there is perhaps more confidence and mastery of the medium, and the best of them certainly show Liszt as one of the outstanding song-writers of the nineteenth century. Liszt wrote no more songs till after 1870; and the songs of the last period, in common with the other works of that time, show, as we shall see, a very radical change of style.

Liszt also embarked at this time on a medium which was fashionable in the drawing-rooms of the day—the melodrama, or poem spoken to music. His first venture in this genre, a setting of Bürger's charnel-house-romantic poem *Lenore* (on which Raff based a whole symphony), is effective and dramatic, but hardly more than that. It dates from 1858 and has piano accompaniment; in the following year Liszt set in the same manner a Schiller centenary tribute by F. Halm (cf. p. 91), *Vor hundert Jahren*, this time with orchestral accompaniment. But a far more remarkable piece is *Der Traurige Mönch*, written in 1860 on a poem by Lenau. The poem is again a typical Gothic-romantic one, but Liszt has transcended the normal melodramatic treatment of such a subject by basing his music almost consistently on the whole-tone scale (cf. pp. 57-8.) The opening passage is shown on p. 97 (Ex. 52).

Liszt himself expressed a fear that " these keyless discords would prove impossible of performance," and one can well imagine how they must have shocked the musicians of the 1860s; but the piece itself is by no means exaggerated or self-consciously experimental. It has in fact been revived with success in modern times, and is certainly one of the most interesting and remarkable works that Liszt ever wrote.

Looking back over the Weimar period, one cannot fail to be struck, not only by the quantity of music that Liszt produced,

but also by the very high quality of a good deal of it. Liszt was at the height of his powers in these years; to his innate poetry and romanticism he had now added the technical knowledge that enabled him to put the most daring experiments into practice, and in addition he was able to concentrate entirely on the work he wanted to do. He came near to realising his aim of making Weimar a really permanent centre of the arts, and as long as he felt this as a possibility he retained his idealistic ardour. Unfortunately many factors, of which the intrigues of his opponents were by no means the only ones, began to dim the bright picture of the future which he had painted for himself, and when he finally left Weimar in 1861 it was in a mood of disillusionment which was to colour his creation for the rest of his life.

CHAPTER IV

THE FINAL PERIOD (1861-86)

PART I — ROME (1861-9)

IT WILL be convenient to divide the final chapter of this survey
into two parts, the first covering the years when Liszt was
mainly living in Rome, and the second the period of his
triangular journeys between Rome, Weimar and Budapest—the
so-called " vie trifurquée." In the first period Liszt's thoughts
turned increasingly towards sacred music, and the external
events of his life no doubt caused him to keep away from the
world which had dealt him so many bitter blows, and seek refuge
in religion. The demonstration at the first performance of
Cornelius' *Barber of Bagdad* in 1858, which Liszt took to be
directed against himself, and which caused him to offer his resignation to the Grand Duke, was followed next year by a tem-
porary breach in his friendship with Wagner. In addition
Princess Sayn-Wittgenstein fell out of favour with the Weimar
court, and Liszt began to feel that it was impossible for him to
remain in Weimar, even in a private capacity. An even more
grievious blow in this year was the death of his son Daniel at the
age of only 21. In 1860 the newspaper protest against the New
German School, of which Liszt and Wagner were regarded as
the leaders, appeared under the signatures of Brahms, Joachim,
Grimm and Scholz; shortly afterwards the Princess left Weimar
for Rome in the hope of obtaining her divorce from the Pope,
and Liszt drew up his last will and testament. Next year he left
Weimar and joined the Princess in Rome; preparations were
made for their marriage, but at the last moment the Pope
revoked his sanction of the divorce. In 1862 Liszt's eldest
daughter Blandine died; in the following year he entered the
Oratory of the Madonna del Rosario at Monte Mario, where he
was visited by the Pope, and in 1865 he received the four minor
orders. The Princess' husband had died in the previous year,
but there was now no further talk of their marriage.

These events find their most direct expression in the *Trois Odes Funèbres* for orchestra, but they are indirectly reflected in other ways, particularly in the number of short religious pieces which Liszt wrote at this time. (He was also working on *St. Elisabeth* and *Christus*, and the Hungarian Coronation Mass was completed during this period too). Most of the sacred works for piano are of no great musical significance. There is an Ave Maria, so-called " The Bells of Rome," an Alleluia and an Ave Maria, the latter transcribed from Arcadelt, two pieces called *A la Chapelle Sixtine*, which are transcriptions of the Miserere of Allegri and Mozart's *Ave Verum* (these also exist in versions for organ and for orchestra) and various smaller pieces. By far the most interesting of all these religious pieces are the two Legends, which are of great beauty and interest. The first is *St. Francis of Assisi preaching to the birds*, and is based on the well-known passage from the *Little Flowers of St. Francis*: " He lifted up his eyes and saw the trees which stood by the wayside filled with a countless multitude of birds; at which he marvelled, and said to his companions: ' Wait a little for me in the road, and I will go and preach to my little brothers the birds.' And he went into the field, and began to preach to the birds that were on the ground; and forthwith those which were in the trees came around him, and not one moved during the whole sermon; nor would they fly away until the Saint had given them his blessing." Liszt says in his preface to the piece: " My lack of ingenuity, and perhaps also the narrow limits of musical expression possible in a work of small dimensions, written for an instrument so lacking in variety of accent and tone colour as the piano, have obliged me to restrain myself and greatly to diminish the wonderful profusion of the text of the ' Sermon to the little birds.' I implore the glorious poor servant of Christ (il glorioso poverello di Cristo) to pardon me for thus impoverishing him." Here Liszt is unduly modest, for the piece is admirably written and an apt illustration of its subject; it contains a theme from Liszt's setting of St. Francis' *Cantico del Sol*.

Even finer musically is the second legend, *St. Francis of Paola walking on the waves*. The story relates that St. Francis was refused admittance to a ferry-boat crossing the Straits of Messina, the boatman remarking : " If he is a saint, let him walk on the water." Whereupon St. Francis spread his cloak on the water; he lifted up part of it like a sail, and supporting this with his staff like a mast, crossed safely over to the other side. The legend is a magnificent descriptive piece; the scene-painting of the foaming waves is brilliantly carried out, and the pictorial and musical elements are admirably fused. The coda is based on a theme from *An den heiligen Franziskus von Paula*, a short work for male chorus.

The other piano works of the period include the well-known concert studies *Waldesrauschen* and *Gnomenreigen*, both admirable works of their kind. Liszt also embarked on a series of technical studies for pianists; these were not completed till about 1880, and were published in twelve volumes after Liszt's death. It seems extraordinary that this exposition of the technique of piano-playing, by one who was perhaps the greatest master of the art who has ever lived, has not come into general use at colleges and academies. Another important piano work is the set of variations on the theme of Bach which is the *basso ostinato* of the first movement of his cantata *Weinen, Klagen, Sorgen, Zagen* and also appears in the *Crucifixus* of the B minor Mass. Liszt had written a short prelude on the same theme a few years previously, and here, as in the Prelude and Fugue on BACH (cf. pp. 87-8), we find the advanced use of chromaticism which ultimately derives from the more experimental passages in Bach. The variations, though beginning more or less in the manner of a normal passacaglia, continue much more freely, and there is a middle section based on figures of this type:

Ex. 53

Clearly the sense of tonality is very strained here. After a return of the main theme, which rises to a climax, the work ends with a statement of the chorale " Was Gott tut das ist wohlgetan." It is a most powerful and impressive piece, and shows a concentrated feeling of gloom and despair such as Liszt did not often achieve with success.*

Apart from the Funeral March in memory of the Emperor Maximilian of Mexico, which will be considered in connection with the rest of the third volume of the *Années de Pèlerinage*, the remainder of the piano pieces of these years are in lighter vein. They include the first two of the five little piano pieces written for Olga von Meyendorff,† the first of which is a reminiscence of the second Liebestraum. The most charming of the set is the third one, an exquisite example of the real simplicity and originality which Liszt could achieve when he wished; this dates from 1873. There is also the *Rhapsodie Espagnole*, subtitled *Folies d'Espagne et Jota Aragonesa*, a much more successful work than the earlier Spanish Rhapsody (cf. p. 42). The first part is a kind of slow passacaglia on La Folia, familiar to us from Corelli and elsewhere, while the brilliant second half contains some of the same themes as the earlier rhapsody. There is also an excellent arrangement of the work for piano and orchestra made by Busoni. The transcriptions include the *Illustrations de l'Africaine* (Meyerbeer's last opera, produced in the year after his death). Liszt composed this at the Villa d'Este almost immediately after he had received the tonsure; and he wrote to the Princess that he was trying out " quelques traits sur le piano pour la jonglerie Indienne de l'Africaine, dont je serai le coq d'Inde, autrement dit, le dindon de la farce." In spite of his new preoccupations his character still remained the same! Other transcriptions include two numbers from Mozart's Requiem, some works of Gounod, and a fantasy on an opera by the Hungarian composer Mosonyi.

The only important orchestral works of the period are the *Trois Odes Funèbres*; but these are extremely interesting, not only for their musical value, but also for the insight they give into Liszt's thoughts at this time. The first, *Les Morts*, was written in 1860 in memory of Liszt's son Daniel. It is described as an " Oration for orchestra with male chorus *ad libitum*." The title is explained by the fact that throughout the score there

---

* It was inspired by the death of his eldest daughter Blandine.

† The fifth piece, *Sospiri*, has only recently been discovered; the MS. is in the Library of Congress, Washington, D.C.

is written a prose passage by Lamennais, which begins as follows: " They too have lived on this earth; they have passed down the river of time; their voices were heard on its banks, and then were heard no more. Where are they now? Who shall tell? But blessed are they who die in the Lord." The last three sentences recur from time to time throughout the work as a kind of refrain, and each time the male chorus enters with the words: " Beati mortui qui in Domino moriuntur." Liszt did not in fact intend Lamennais' words to be declaimed during the music—they were simply inserted in the score as a guide to the musical thought—but the work has been so performed in modern times with very moving effect. It is a very fine and dignified elegy, rising to a great central climax with the words " Holy, holy, holy, is the Lord God of Hosts," and afterwards returning to the quiet mood of the opening. Liszt clearly had a special affection for it, and in his will asked Princess Sayn-Wittgenstein to see that it was published; but in fact it did not appear till many years after his death, in the Breitkopf Collected Edition.

The second Ode, *La Notte*, is even more interesting. It was written in 1864, two years after the death of Liszt's eldest daughter Blandine; the main part of it is an orchestrated version of *Il Penseroso* from the Italian book of the *Années de Pèlerinage* (cf. pp. 30-1), Liszt prefixed to *La Notte*, as he did also to *Il Penseroso*, the quatrain of Michelangelo, " Grato m'è il sonno." These words are significant of Liszt's state of mind at the time; but, even more significantly, he wrote in *La Notte* a new middle section of which a notable feature is the well-known " Hungarian cadence," and to this he prefixed a quota-

Ex. 54

tion from Virgil: " Dulces moriens reminiscitur Argos." (" Dying, he remembers fair Argos.") These words refer to the death in Italy of Antores, an Argive companion of Aeneas, who was killed in battle by a spear aimed at his leader; Liszt, who was living in Rome at the time, clearly felt that he too

might die far from his native Hungary. Liszt asked in a note on
the score that *Les Morts* and *La Notte* should be played at his
own funeral; but his wish was not fulfilled, and both works
remained unperformed until 1912.

The third of the Funeral Odes, *Le Triomphe Funèbre du Tasse*,
also has a personal significance. It was written as an epilogue
to the symphonic poem *Tasso*, and is based on some of the same
themes. Tasso, like Liszt, won a brilliant success at an early
age, but thereafter found recognition withheld from him; and
this work clearly symbolises the idea that Liszt's true fame, like
Tasso's, would not come about until after his death—here
Liszt showed himself a true prophet, for it is only in recent
years that the real significance of Liszt's later works has begun
to become apparent. This is the only one of the three Odes to be
published and performed in Liszt's lifetime; the first performance took place in 1877 in New York under Leopold Damrosch,
to whom Liszt dedicated the work. Harmonically the Odes do
not show any great difference in style from the works of the
Weimar period; but the atmosphere is more dignified and
restrained, and they are fine musical works as well as moving
human documents. They are all published in the Collected
Edition, and well merit revival to-day.

The organ works of this period are of no particular
importance; though fairly numerous, they are mainly
transcriptions of works written for other media. There are,
however, a couple of curiosities in the shape of two transcriptions for trombone and organ, made for Eduard Grosse, the
trombonist of the Weimar Court orchestra; Liszt seems to have
been deeply attached to Grosse, and remembered him in his
will. The first of these transcriptions is a Hosannah arranged
from the *Cantico del Sol* (cf. p. 107); the second is an arrangement of the bouncing tenor aria, " *Cujus animam*," from
Rossini's Stabat Mater, complete with cadenza going up to
high D flat. Tastes change with the times, and recently this
piece found itself perfectly at home in a semi-comic series
devised by the B.B.C. Third Programme entitled " Musical
Curiosities."

The Legend of St. Elisabeth was suggested by the installation in
1855 in the Wartburg, the ancient castle of the kings of
Thuringia near Eisenach, of some frescoes by Moritz von

Schwind representing various episodes in the life of St.
Elisabeth.   Elisabeth, the daughter of King András II of
Hungary, was born in 1207, and was brought to the Wartburg
at the age of four as the future bride of Ludwig, the son of
Landgrave Hermann of Thuringia.   They were married in
1220; Elisabeth's many acts of charity to the poor at first
aroused the anger of her husband, but he was converted by
the Miracle of the Roses.   Later, after Ludwig had been killed
in a Crusade, his mother Landgravine Sophie banished
Elisabeth and her children from the Wartburg, and she lived
in poverty for a time.   On being reinstated, she renounced her
rights in favour of her son; she died in 1231 and was canonised
four years later.

Liszt's oratorio is in six scenes, each representing one of
Moritz von Schwind's frescoes.   The text was written by Otto
Roquette, a professor at the Darmstadt Polytechnikum; Liszt
received the first numbers in 1856, and completed the work in
1862.   Before the first scene there is an introduction which is
mainly based on a plain-chant usually sung on the feast of
St. Elisabeth;   Liszt uses this theme throughout the work, in
the manner of a *leit-motiv*, to represent Elisabeth.

The first scene shows the arrival of the child Elisabeth at the
Wartburg; it is mainly a ceremonial scene, with songs and
choruses of welcome.   In the solo of the Hungarian magnate
appears a Hungarian national theme which is used as a symbol
of Hungary throughout the work.

In the second scene Elisabeth and Ludwig are now married,
and Ludwig has succeeded to his father's throne.   There has
been a famine in Thuringia, and Elisabeth has sold most of her
possessions in order to relieve the distress of the poor; this
arouses the anger of Ludwig's mother, and Elisabeth has to do

her deeds of charity in secret. Ludwig returns home un-
expectedly from one of his campaigns, and meets Elisabeth in
the forest carrying a basket of bread and wine for the poor.
She pretends to him that her basket is full of flowers that she has
been picking, but Ludwig makes her confess what she is really
doing; at this moment the bread and wine are miraculously
transformed into roses. Both are astonished, and Ludwig asks
Elisabeth's forgiveness.

The third scene shows the departure of the Crusaders for
Palestine, with Ludwig at their head. This is again mainly
a ceremonial scene, though there is a short interlude in which
Ludwig and Elisabeth say farewell to each other. At the end of
the scene comes the well-known Crusaders' March; the music
is dominated by the three-note figure which Liszt used to
symbolise the Cross (cf. p. 92).

In the fourth scene Landgravine Sophie has heard that
Ludwig has been killed in battle; she determines to drive
Elisabeth from the Wartburg and regain power over the land,
and in spite of her entreaties, Elisabeth and her children are
turned out of the castle in the middle of a raging thunderstorm.
When the storm abates, Elisabeth has reached (in the fifth
scene) the shelter of a hospital which she had founded for the
relief of the poor; a chorus of poor people praise her for her
good deeds with an old Hungarian " Hymn to St. Elisabeth."

Ex. 57

The scene ends with the death of Elisabeth; an orchestral
interlude, based on the main themes of the work, leads to the
final scene, the solemn ceremony of Elisabeth's interment—
another massive ceremonial scene, led by the Emperor
Friedrich II.

St. Elisabeth is thus partly oratorio, partly drama; and to
perform it on the stage is obviously absurd. The music contains
some striking moments, particularly in the more dramatic
scenes such as the fourth; but it is nowhere much more than

competent and effective. The work, however, does have continuity, and is able to bear its length without undue strain; but one can hardly call it one of Liszt's masterpieces. It was first performed in Hungarian at Budapest in 1865, and was at once hailed as the foundation of a new national Hungarian school; but the music still has a distinctly cosmopolitan flavour, in spite of the use of Hungarian themes here and there, and certainly provides no parallel to the work of Smetana in Bohemia or of Balakireff in Russia. It was perhaps something relatively new at the time, but it has since been surpassed in its genre.

Liszt's next oratorio, *Christus*, is a different matter; for here he had no need to write " effective " ceremonial scenes, and was able instead to express his own reaction to the Bible story. He first had the idea of the oratorio as far back as 1853, and toyed with the idea of inviting various collaborators—Herwegh, the Princess and Cornelius—to prepare the text; but in the end he made it himself, using passages from the Bible, the Catholic Liturgy and some Latin hymns. It falls into three parts, the first being a Christmas Oratorio, the second representing various incidents in the life of Christ, and the third being devoted to the Passion and Resurrection. The first part begins with an introduction which leads straight into a pastorale, with the apparition of the herald angels; then follows a hymn, *Stabat Mater speciosa*, and two orchestral pieces—the Song of the Shepherds at the Manger and the March of the three Holy Kings. In the second part, called After Epiphany, we have first the Beatitudes, followed by a Pater Noster; both had been written some years earlier, and were actually performed as separate pieces before *Christus* was completed. The next piece, The Foundation of the Church (" Tu es Petrus "), had also been composed earlier, and was originally published as *Pio IX*, *Der Papsthymnus*. Then follows a magnificent orchestral piece with some short vocal interludes, The Miracle of the Storm. The second part ends with a description of the Entry into Jerusalem, with choral Hosannahs.

The third part begins with " Tristis est anima mea " for baritone and orchestra, in which the orchestra as it were comments on Christ's words; then follows a long choral setting of the *Stabat Mater dolorosa*. A short Easter hymn,

O filii et filiae, for children's voices and organ, leads to a final triumphant *Resurrexit*. Though *Christus* is in many ways a patchy work, it does represent a self-summing-up of the kind that we find in the piano sonata and the first movement of the Faust Symphony; Liszt combines his varied elements, ranging from Gregorian plain-chant to romantic and dramatic orchestral colour, with consummate mastery, and throughout the work one feels that he sincerely meant every note of it. It is certainly the most successful of his larger choral works, and deserves to be revived.

The most important of the other sacred choral works of this period is the Hungarian Coronation Mass, written for the Coronation of Franz Josef as King of Hungary in 1867. Here again we have an effective ceremonial piece, with many dramatic passages, but not a work of very great musical importance. The Offertorium and Graduale (Psalm 116) were composed after the first performance of the Mass and added to it later. Of the other choral works it is only necessary to mention the *Cantico del Sol* of St. Francis of Assisi (cf. pp. 99, 103) for baritone solo, male chorus, organ and orchestra, and two short liturgical masses, the Missa Choralis of 1865, which makes a considerable use of Gregorian themes, and the male voice Requiem of 1867-8. The latter contains some remarkable whole-tone effects, such as the following:

Liszt also composed a number of shorter religious works at this time, of no great musical importance. No songs or recitations date from this period, and we may therefore now proceed to the consideration of the works of Liszt's last years, many of which showed a radical change in his harmonic style and are of prophetic interest to later generations.

PART II—ROME, WEIMAR, BUDAPEST (1869-86)

MOST of the works composed in the last fifteen years of Liszt's life, when he had emerged from his semi-retirement in Rome and was travelling regularly between his three main centres of residence, together with many visits to other cities, show some quite remarkable innovations compared with the majority of the works so far discussed. The style has become extremely stark and austere, there are long passages in single notes and a considerable use of whole-tone chords, and anything resembling a cadence is avoided; in fact, if a work does end with a common chord it is more often in an inversion than in root position. The result is a curiously indefinite feeling, as if Liszt was launching out into a new world of whose possibilities he was not quite sure. For the majority of these works he returned to his first love, the piano; but in general the old pianistic glitter is absent—Liszt was now writing for himself, and no longer for his public. In fact, a good many of these pieces were not even published until many years after Liszt's death, and some are still in MS. to-day.

The principal collection of late piano pieces is the third year of the *Années de Pèlerinage*, written at various times between 1867 and 1877. The volume consists of the following pieces:

1. Angelus—Prière aux anges gardiens.
2. Aux Cyprès de la Villa d'Este I (3/4).
3. Aux Cyprès de la Villa d'Este II (4/4).
4. Les jeux d'eaux à la Villa d'Este.
5. Sunt lacrymae rerum, en mode hongrois.
6. Marche funèbre.
7. Sursum corda.

Of these Nos. 1 and 7, the definitely religious pieces, are the most conventional: *Angelus*, which also exists in versions for harmonium and for string quartet, is pleasant but of no great distinction, while *Sursum corda* bears a certain resemblance to the *Invocation* from the *Harmonies Poétiques et Religieuses*, though the harmonic style is more advanced and contains some use of the whole-tone scale.

The earliest piece in the collection is the Funeral March in memory of the Emperor Maximilian of Mexico, who was executed in 1867. This is a curious, somewhat shapeless, but interesting piece; and so is " Sunt lacrymae rerum," written with some use of the Hungarian scale. The finest pieces in the collection are the three associated with the Villa d'Este, where Liszt had the use of a suite of rooms, and spent part of each year at this time. Of these, *Les jeux d'eaux à la Villa d'Este* has become fairly well known; its impressionistic use of diatonic harmonies had a considerable effect on Ravel when he came to write his *Jeux d'eaux*, and this piece in turn influenced the French impressionist school of the period. These are the opening bars of Liszt's piece:

Ex. 59

Less well known, but even finer musically, are the two Threnodies, *Aux Cyprès de la Villa d'Este*, which give an extraordinary atmospheric feeling. The second, in 4/4 time, is perhaps the better of the two; its opening bears a curious resemblance to the beginning of *Tristan* (cf. p. 52):

Ex. 60

The *Christmas Tree* suite, written between 1874 and 1876, is a mixed collection of religious and genre pieces. They are mainly simple in style, and some of them are extremely charming, such as the setting of " In dulci jubilo " and the little Scherzoso, " Lighting the candles on the Christmas Tree." There are two bell pieces, *Carillon* and *Cloches du soir*, and the collection ends with three secular pieces, a nostalgic waltz called *Jadis*, a rather sinister little march in the Hungarian style, and a Mazurka which alternates between nostalgia and gaiety.* These few bars from *Jadis* illustrate its rather curious atmosphere of looking back into the past—a feeling which is found in many of these late works of Liszt.

Ex. 61

There is also a third collection, of seven Historical Hungarian Portraits—tributes to various Hungarian political and artistic leaders. These were first published in 1956 in Budapest. The fourth piece, " Ladislaus Teleky," is identical with the Funeral March discussed on p. 113, though without its Prelude. The first, second, third and fifth pieces are in Liszt's Hungarian style—fiery and effective, but not startlingly original. The first five pieces in the collection were written in 1885; the sixth and seventh, of an earlier date and previously published separately, were written in memory of the composer Mosonyi

* The two latter are portraits of Liszt himself and the Princess respectively—*Jadis* recalls their first meeting.

and the poet Petőfi, respectively.   The Mosonyi piece is a fine,
grave elegy, which deserves to be established in the repertoire;
the Petőfi elegy is slighter, and is based on a theme from Liszt's
recitation " Des toten Dichters Liebe " (" the dead poet's
love ")—a setting of a poem by Moritz Jókai; Petőfi was only
twenty-six when he was killed in the 1849 Revolution.   Both
these pieces are fairly simple in style.   Liszt also made an
orchestral version of the whole set, which seems to have been
lost.

Several other Hungarian works also date from this period.
These include the Hungarian Rhapsodies Nos. 16-19, which are
considerably more experimental and less brilliant and superficial than their predecessors, though they do not perhaps quite
achieve the state of being really satisfactory pieces owing to a
certain fragmentariness of treatment.   There are also some
simple arrangements of five Hungarian folk songs, which are
interesting in that they anticipate Bartók's work in the same
field.   The second one of these is particularly charming.

Ex. 62

There are also three Csárdás, of which the most remarkable are
the *Csárdás Obstiné* and the *Csárdás Macabre*. The latter is indeed
extraordinary for its period in that it is mostly written in bare
parallel fifths; here is part of the opening section.

Ex. 63

Its ending rises to a violent and furious climax. (Incidentally, the theme of the two Trios is similar to a Hungarian folk song in Bartók's collection.)*

Many of the piano works of this period are short pieces which express a particular mood. There is, for instance, the charming nocturne *En Rêve*, a rather more disturbed nocturne called " Sleepless, Question and Answer," and two Elegies, of which the first was originally written for cello, piano, harp and harmonium. It was in fact a characteristic of Liszt at this time to write pieces which could be played on either a chamber combination or for piano solo;   the music is, as it were, " abstract," and depends very little on instrumental colour, and probably many of these pieces were written down for piano more for the sake of convenience than anything else. We can see this in the little group of four pieces which are connected with the death of Wagner: the two versions of *La lugubre gondola, Richard Wagner—Venezia,* and *Am Grabe Richard Wagners.* Of these *La lugubre gondola* I and II were written in Venice about two months before Wagner died there; they were inspired by the funeral processions by gondola which Liszt saw on the lagoons — and it is certainly a most curious coincidence that the body of his lifelong friend should have been carried in this way so soon afterwards. Both versions have a simplicity and austerity which remind one of Bartók; here is part of the first version:

The second version was originally written for cello or violin and piano; and *Am Grabe Richard Wagners* is scored for string quartet and harp, with piano solo as an alternative. It is a short and simple piece, and is prefaced with the following note:

* A slightly longer version of this work was recently discovered and published in Budapest in 1955.

" Wagner once reminded me of the likeness between his Parsifal theme and my previously written *Excelsior!* [Introduction to *The Bells of Strasbourg*—a setting of Longfellow's "Golden Legend."] May this remembrance remain here. He has fulfilled the great and sublime in the art of the present day. F. Liszt. 22 May 83. Weimar." * This piece, which is partly based on the " bell " theme from *Parsifal,* has recently been published for the first time by the Liszt Society.†

Three of these pieces show Liszt's harmonic experimentation at its most extreme. The first is a Funeral Prelude and March, (cf. pp. 110-1) written in 1885, of which the March is based on an ostinato figure of four notes—F sharp, G, B flat, C sharp; these are driven against the accompanying harmonies with the greatest violence, leading to clashes of this kind:

Ex. 65

Similarly at the climax of *Unstern* (Sinistre) we find the following passage:

Ex. 66

An even more remarkable example is the short piece *Nuages gris,*

* The date is, of course, that of Wagner's 70th birthday; on that day Liszt conducted in a Wagner memorial concert at Weimar.

† The opening theme of the piece is common to both *Excelsior!* and *Parsifal.* An interesting account of the relation between the three works and of the psychological significance of this theme for Wagner will be found in " Liszt and *Parsifal* " by Arthur W. Marget (Music Review, Vol. XIV, No. 2, May 1953).

an extraordinary example of impressionism. In the opening
section the blurred effect of the pedal is deliberately marked by
Liszt, and in the final passage the chromatic rising phrase is
driven against the whole-tone harmonies in the left hand without any regard for orthodox consonance.

One would certainly hardly believe that this piece was written
by the composer of brilliant fantasies on themes by Rossini and
Donizetti.

A few more of the late piano works deserve mention; many
of them, like the piano piece in F sharp, *Recueillement*, and the
Impromptu are simple and charming, while others, like the
Romance Oubliée, have a distinctly nostalgic flavour, comparable
with that of *Jadis*. The title *Romance Oubliée*, incidentally,
derives from the fact that this is a new version of a Romance
written more than thirty years earlier*; in the same way Liszt
wrote four *Valses Oubliées* at this time which are also nostalgic
evocations of the past.† The first of these four waltzes is the
only one of Liszt's late piano works accepted into the general
repertoire; but most pianists completely destroy its essential
character by playing it much too fast. Very different in feeling
are the last three Mephisto Waltzes and the Mephisto Polka.
The second waltz was originally written for orchestra and will
be discussed below, and the Mephisto Polka is overlong and not
of great interest. But the third Mephisto Waltz is certainly

* For the curious history of this piece, see Friedrich Schnapp: "A Forgotten
Romance of Liszt" (Music and Letters, Vol. XXXIV, No. 3, July 1953).

† The fourth waltz was discovered in the U.S. and published there in 1954 by
Theodore Presser Co.

one of Liszt's finest achievements. It begins with this startling phrase:

The mood is angry and violent throughout, with little relief; it shows Liszt at his most ruthless and savage. The fourth Mephisto Waltz was not revised for publication by Liszt, and in fact has only recently been published by the Liszt Society. It is complete as it stands in the sense that it is possible to play it through from beginning to end; but Liszt also left some sketches for a contrasting section which he intended to interpolate towards the end, and so one cannot speak of the work as properly finished. In its present form the work is not really up to the level of its three companions, but had Liszt lived to revise it he might well have made considerable improvements.

A number of other late works have only come to light in recent years: they include the fourth *Valse Oubliée*, of which the MS. was given by Liszt to a pupil who later settled in the U.S.; a previously unknown Toccata of 1879, of which the MS. is now in the Library of Congress, Washington, D.C.; and—most remarkable of all—a *Bagatelle sans Tonalité*, which was found in the Liszt Museum at Weimar and published in Budapest in 1956. Written in 1885, this was originally intended to be the fourth Mephisto Waltz. Though it is not atonal in the Schoenbergian sense, it certainly lacks any definite key feeling, being mainly based on tritone and diminished seventh harmony, and ending in a curiously indefinite way.

The piano transcriptions of these final years are varied and fairly numerous. Apart from transcriptions of a number of works by unimportant and now forgotten composers, Liszt arranged for piano Bach's organ Fantasy and Fugue in G minor, and Saint-Saëns' *Danse Macabre* (a great improvement on the original work!); the transcription of the Sarabande and Chaconne from Handel's *Almira* amounts to an original work on Handel's themes. His interest in the new school of Russian nationalist composers, which he did so much to help in their early years, is shown in the transcriptions of a Tarantella by Dargomijsky, originally written for three hands, of which one plays a constant tic-tac bass on A, and of another Tarantella

by César Cui; he also transcribed the Polonaise from
Tchaikovsky's *Eugen Onegin*, in spite of Tchaikovsky's un-
complimentary opinions of him! He continued too, with the
series of Wagner and Verdi transcriptions mentioned above
(cf. p. 63); his last operatic fantasy, that on Verdi's *Simone
Boccanegra*, is a fine, dignified work, quite different in style from
the early brilliant fantasies, and containing some remarkable
whole-tone harmonies in the middle section. The transcription
of the March to the Grail from *Parsifal*, made in 1882, also
shows the same dignity and restraint.

We have seen that the first Elegy was originally written for
cello, piano, harp and harmonium; and similarly other piano
works of this period also exist in versions for various chamber
combinations. There is also one work especially written for
violin and piano, the *Epithalam* composed for the wedding of
the violinist Reményi in 1872; and another work, *Die Wiege*, for
four violins would appear to have some connection with the
first part of the thirteenth and last symphonic poem, *From the
Cradle to the Grave* (1881-2).

This was inspired by what is apparently a very bad painting by
Count Michael Zichy, and is divided into three parts, " The
Cradle," " The Struggle for Existence," and " To the Tomb :
the Cradle of the Future Life." The scheme suggests the worst
type of Victorian melodrama; but the originality of the music
makes it a very interesting work. The first part is tender,
delicate and beautifully written; the violent central section
is admittedly rather more conventional, but at any rate acts
as an effective foil to the outside parts; while the final part
combines the themes of the two previous sections and uses them
so as to suggest a curious, transitional atmosphere which is
exactly in keeping with the subject. This short extract will
perhaps give some idea of the last section:

The second Mephisto Waltz, for orchestra, was written in 1880-1, and is dedicated to Saint-Saëns—a rather doubtful compliment when one considers how much more violent its expression is than that of Saint-Saëns' own *Danse Macabre*, which, as we have seen, Liszt had himself transcribed for piano. It is a most powerful and effective piece, with a characteristic ending; after building up a big climax in E flat, the main key throughout most of the waltz, the music suddenly falls on to the tritone B natural—F, and so ends the piece in an entirely unexpected and startling manner.

The only other important work for orchestra is *Salve Polonia*. This was written as far back as 1863, and may even have been sketched as early as 1850; it was originally intended to be a separate piece, but Liszt later inserted it as an interlude in his unfinished oratorio *St. Stanislaus*. Liszt got the idea of writing an oratorio on the life and death of the Polish saint from the poem by Lucian Siemienski, which was (very indifferently) translated into German by Cornelius from a French version, made apparently by the Princess. Liszt was not very happy with the text, and made frequent attempts to get it improved by various writers; but it was not till 1883 that this was finally done by K. E. Edler, who incorporated some new material discovered in a chronicle at the monastery of Ossiach, near Klagenfurt. Liszt worked at the music intermittently up to the end of his life, but it was left unfinished at his death. The original plan was for a dramatic oratorio in six scenes somewhat on the lines of *St. Elisabeth* (those interested in the details will find them set out in Raabe II, p. 142-3); as a result of the new material discovered in 1883 Liszt altered his projected ending by including the Psalm *De Profundis* (No. 129) which he had written two years earlier, and also by using *Salve Polonia*, not as an interlude, but to end the whole work in a new version with solo baritone and chorus. The whereabouts of the MSS. of the greater part of the work are not known; but they are unlikely to have been lost, and will no doubt see the light of day at some time. *Salve Polonia* was published in 1883, and the Liszt Museum also possesses MSS. of part of the first scene and of a piano transcription of two Polonaises from the work.

Liszt wrote a number of shorter sacred choral works during this period, many of them designed for liturgical use. As we have seen, his aim was to produce a new type of church music which would be more expressive and poetical than the academic productions of his contemporaries; but, as may be imagined, he met with considerable resistance from official bodies. His large-scale works for chorus and orchestra did indeed meet with a certain amount of success, perhaps because it was possible to perform them under normal concert conditions; but his attempt to revolutionise liturgical music was a failure. This may have been Liszt's own fault, for on the whole these works do not show him at his most original; but there are some striking things among them the fifty or so works of this kind which he composed. " Qui seminant in lacrimis " contain some curious harmony, and in particular the opening section

of the motet " Ossa arida—O ye dry bones, hear ye the word of the Lord " is entirely based on a chord which gradually piles up all the notes of the diatonic scale in the organ part (which incidentally requires two players); at the entry of the male chorus the whole chord is sounded fortissimo—a truly terrifying effect.

But the most remarkable of all these late choral works is

Via Crucis—the fourteen Stations of the Cross for soloists, chorus and organ, completed in 1879. The text was arranged by Princess Sayn-Wittgenstein from Biblical quotations, Latin hymns and German chorales, and soloists represent Jesus, Pilate, and the mourning women. The whole work has a most curious atmosphere, for while it is entirely restrained and devout in feeling, there is a more or less consistent use of experimental harmony, particularly that derived from the whole-tone scale. This can be seen in passages such as these:

Via Crucis does, in fact, represent the fulfilment of Liszt's aim to create a new kind of church music by allying a new harmonic technique to the old liturgical framework; he clearly had the subject very much at heart, and the result is not a mere experiment, but a very deeply felt and moving work. To show how little Liszt's new style was understood by his contemporaries it must be recorded that *Via Crucis* was rejected by a prominent firm of publishers of religious music in Liszt's lifetime, and was not in fact performed or published till more than forty years after his death. Yet after the first public performance in London, which took place in 1952, a critic wrote: " Now and again, of course [in this work], Liszt achieved what he wanted by more or less orthodox (but strikingly beautiful) nineteenth-century means. . . . Yet for the greater part of *Via Crucis* Liszt attained extreme expressiveness, not by organic development or contrapuntal complexity, but by a forthright ' primitivity ' remarkably attuned to the twentieth-century spirit and ideal." Liszt said: " I can wait "; and it is certainly true that the real significance of his final period has only become apparent to us in recent years. (Incidentally the majority of Liszt's shorter choral works, as well as some of the Masses, are published in the Breitkopf Collected Edition.)

The organ music and secular choral works of this period are of no great importance,* and we may conclude this survey with a discussion of the late songs. These are extremely interesting, for they show the same tendencies as we have noted in the piano music—austerity, restraint, even inconclusiveness at times. There are about fifteen of them, written at various times from 1871 onwards; the title of one of them, de Musset's " J'ai perdu ma force et ma vie," perhaps symbolises the general mood of resignation which we find in them. There is certainly nothing which could be called " effective " vocal writing in the manner of some of the early songs. One can see this in Liszt's only English song, a setting of " Go not, happy day," written in 1879 for a Tennyson album; its very quiet and restrained character is in marked contrast to some of the better-known settings by native composers. In *Verlassen*, which was written to be sung by an actress in a play, the vocal compass is kept extremely small, and the emotional feeling of the song is chiefly brought out in recitative-like passages. In many of these songs, in fact, we can see Liszt approaching the intimate quality of Hugo Wolf, and turning his back on full-blooded romanticism. There is, however, one song of a rather more dramatic nature, in spite of its extreme brevity—*Und wir dachten der Toten*, a setting of the final verse from a poem by

* Apart from the interesting " Missa pro organo."

Freiligrath. The final passage is worth quoting as a further example of Liszt's experimental style:

Ex. 73

Liszt was an extrovert composer who threw out his ideas in all directions. Though he frequently revised and rewrote his works both before and after publication, one does not find with him the long and laborious process of sketching and polishing by which Beethoven arrived at the final versions of his major works. Consequently, composition was often a hit-or-miss process with Liszt; in all periods of his life he probably wrote as many works of no particular musical importance as those which, I have suggested, are worthy of our continued interest and attention. One may not find that his works are always completely finished and unalterable in the way that those of Mozart and Beethoven are; but what is important with Liszt is not only the number of ideas that came to him, but also the remarkable quality of very many of them. He certainly lacked self-criticism on many occasions; but he always felt a burning necessity to express the ideas which came into his mind. And there is no doubt that in this last period he did become the prophet of the music of later generations, not so much through his harmonic innovations alone as through his general approach to music. The romantic fervour of which Liszt was one of the chief apostles in the 1830s had by this time been replaced on the one side by the more classical approach of Schumann and Brahms, and on the other by the monumental dramatic efforts of Wagner and his followers. As we in this century have seen, both these roads turned out in the end to be *culs-de-sac*; the Brahms tradition led only to minor figures like Dohnányi and Medtner (though admittedly Schoenberg in his own way learnt something from it); the Wagnerian colossus blew itself up and finally exploded with Strauss and Mahler. Liszt, in spite of his enormous admiration for Wagner, must have seen this; for why else should there have been such a radical change in his style in these years? In fact the road on which he started is that which a large number of composers have adopted to-day, a style in which every note is of importance and nothing is wasted or put in merely for effect. It is a style in which the feeling of key is left deliberately vague; though it may contain impressionistic elements on occasions, it is not itself fundamentally impressionist; it simply aims at the expression of a mood or an idea in the most direct and basic form. One can see how alien this approach must have been to the musicians of the late nineteenth century; and this may explain why so many of these pieces remained unpublished— and even those which were published at the time were misunderstood and regarded as mere senile scribblings. One cannot but admire the courage of the ageing Liszt in striking out into new and uncharted ways at a time of life when he could well have rested on his laurels, and, in spite of complete lack of appreciation, taking the path which he knew the music of the future must follow.

Liszt was born two years after the death of Haydn; he died a year after the birth of Alban Berg, and in a way he may be said to bridge the gap between those two musical words. When one thinks of the long road that he had to travel between his early Czernian pieces and late works like *Nuages gris* and *Unstern*, his achievement becomes even more astonishing, Nor can one regard him simply as a technical innovator whose works have no actual musical value apart from the experiments contained in them. Each work has its own idea, mood, atmosphere or emotion to express, and each work presents a genuine musical experience. Admittedly many of Liszt's works are superficial, overwritten or merely dull—in so large an output one could hardly expect otherwise, and the same may be said of many other great composers. But the point at issue is that he did also write a great deal of extremely fine music, and it is by that that he should be judged—nobody would dream of assessing Beethoven's powers on the strength of the Battle Symphony alone. One may dislike the flavour and quality of some of Liszt's music, but that is purely a matter of taste, and cannot possibly be made into a universal judgment. Liszt was a person who was torn in many directions simultaneously, and this may be seen in the varied style and quality of his music. But above all he did sincerely feel whatever he wrote at the time he wrote it; and it is that which will make his music live. No doubt he will always remain a controversial figure, just as Berlioz has done; but we must salute him for his unique contribution to the music of his time, and must also reflect that without that contribution the music of our time too would be very different.

BIOGRAPHICAL SURVEY

THIS SURVEY attempts to present the main events in the life of Liszt—not in very great detail, for which readers must be referred to the numerous biographies—but sufficiently to show the principal course of his life. In parallel with this is a list of his works, arranged as far as possible in order of composition. The dates given here for some of them appear somewhat dogmatic; in many cases there is considerable uncertainty about these, and some works were, of course, completed over a period of several years. More exact details will be found in the catalogue at the end of this book; the intention here is merely to indicate the relation between his life and his works in a more or less general way. Dates of publication are only given in cases where the date of composition or transcription are unknown, except for collections of pieces, like the *Années de Pèlerinage*, for instance. The dates of important first performances of Liszt's works are given, but normally not those of subsequent performances. Arrangements by Liszt of his own works for other media are not mentioned in this chapter, apart from the few that have become famous, e.g., the Petrarch Sonnets and the *Liebesträume*.

Year	*Life*	*Work*
1811	*October 22:* Liszt born at Raiding, Government of Sopron, Hungary. Son of Adam Liszt, an official in the service of Prince Nicolas Eszterházy.	
1817	Begins to listen attentively to his father's piano playing and to show interest in church and gipsy music. Develops a strong religious sense.	
1818	His father begins to teach him the piano, and he makes astonishing progress.	

Year	*Life*	*Work*

1819 Begins to compose in an elementary way. He plays the piano well and reads easily at sight.

1820 *October:* Plays for the first time at a public concert at Sopron.
November 26: Plays at a second concert at Poszony in the palace of Prince Eszterházy; a number of Hungarian magnates subscribe towards the expenses of his musical education.

1821 Moves with his family to Vienna, and studies with Czerny (piano) and Salieri (composition).

1822 Great progress made. Success in Austrian and Hungarian aristocratic circles. Meeting with Schubert.
December 1: First public concert in Vienna.

Tantum ergo (lost).
Variation on a waltz of Diabelli.

1823 Many public and private engagements.
Visit to Beethoven.
April 13: Second concert in Vienna.
Autumn: Moves to Paris with family. Concerts in many German towns on the way, including Munich and Stuttgart.
December: Arrival in Paris. Refused admission to the Conservatoire by Cherubini on the ground that foreigners are not allowed. Studies composition with Paër.

1824 Makes a success in Parisian society and plays at many fashionable concerts.

Year	*Life*	*Work*
1824	*March 7:* First public concert in Paris, at Theatre Louvois. *May:* First visit to England. Plays at many private concerts and at court. *June 21:* First concert in London.	Impromptu sur des thèmes de Rossini et de Spontini. Huit Variations. Variations brillantes sur un air de Rossini. Allegro di bravura. Rondo di bravura.
1825	*Spring:* Tour through the French provinces. *June:* Second visit to England. Concerts at Manchester, at Windsor Castle before George IV, and at Drury Lane Theatre in London. *June 16:* New Grand Overture (that to Don Sanche) first performed at Manchester. *October 17:* Don Sanche first performed at the Opéra, Paris.	Don Sanche, ou le Château de l'Amour, opera.
1826	Second concert tour through the French provinces. Study of counterpoint with Reicha. *Winter:* Tour in Switzerland (Geneva, Lucerne).	Étude en douze exercices.
1827	*Early spring:* Return to Paris. *May:* Third visit to England. *June 9:* Piano Concerto played at a concert in London. L. suffers from nervous exhaustion and wishes to become a priest. Taken to Boulogne by his father, who dies there. *August 28:* Returns to Paris and teaches the piano.	Scherzo in G minor. Piano Concerto (forerunner of the Malediction): lost, except for a fragment.
1828	Falls in love with an aristocratic pupil, Caroline, daughter of Comte de Saint-Cricq, who insists that the attachment be broken off.	*May 21:* Zum Andenken.

Year	*Life*	*Work*

1828   L. again wishes to join the Church, but is dissuaded by his mother. He becomes ill, *winter*, and rumours of his death begin to circulate.

1829   Slow recovery from illness; period of religious doubts and pessimism, followed by return to public life.

           Work: Grande Fantaisie sur la Tyrolienne de l'opéra La Fiancée (Auber).

1830   Takes a great dislike to the career of a virtuoso. Reads widely to supplement his defective education and meets many artists, including Lamartine, Victor Hugo and Heine. Becomes a supporter of the Society of Saint-Simon. *December 4:* Meets Berlioz for the first time, and attends the first performance of his Symphonie Fantastique on the following day.

           Work: July: Sketch of Revolutionary Symphony.

1831   *March 9:* Hears Paganini for the first time, and again becomes interested in virtuoso technique. Meeting with Chopin and Mendelssohn.

1832   Friendship with Chopin develops. L. hears him play in public on *February 26*, and becomes an excellent interpreter of his music. Goes to Fétis' lectures on the philosophy of music. Visit to Savoy.

           Work: Grande Fantaisie sur la Clochette de Paganini.

1833   *April 22* and *December 2:* Plays in two concerts arranged by Berlioz. *October 3:* Witness at Berlioz' wedding to Harriet Smithson. Seeks religious instruction from the Abbé Lamennais.

           Work: Transcribed: Symphonie Fantastique of Berlioz. Overture Les Francs-Juges of Berlioz. L'Idée Fixe.

Year	*Life*	*Work*
1834	Through Alfred de Musset L. meets George Sand and the Comtesse Marie d'Agoult; love affair begins with the latter. *Summer:* Prolonged visit to the Abbé Lamennais at La Chesnaie in Brittany. Essay " On the Future of Church Music." *November 24:* L. gives first performance of the Lélio Fantasy in a Berlioz concert; also the Konzertstuck on themes of Mendelssohn.	Grande Fantaisie Symphonique on themes from Lelio. Grosses Konzertstuck über Mendelssohns Lieder ohne Worte. Harmonies poétiques et religieuses (single piece). Apparitions. De Profundis. Lyon.
1835	*May:* L. plays in another Berlioz concert. Marie d'Agoult leaves her husband and family for L. and they join each other in Switzerland. *December 18:* L.'s first daughter, Blandine, b. Geneva. L. becomes a teacher at the newly founded Conservatorium in Geneva. Essays " On the Position of Artists " appear in the Gazette Musicale.	Fantasies on La Juive, Niobe and Lucia. Fantaisie romantique sur deux mélodies suisses. La pastorella dell' alpi e Li marinari. La Serenata e L'Orgia. Duo, Le Marin. Album d'un voyageur. La Rose (Schubert).
1836	*Spring:* Plays in Lyon. Returns temporarily to Paris to assert himself against the growing success of Thalberg. *Summer:* Returns to Geneva. *December 18:* Plays at a Berlioz concert in Paris, and remains there for some months, together with Marie d'Agoult.	Rondo fantastique, El Contrabandista. Fantasies on Les Huguenots and I Puritani. Transcriptions of Berlioz' Harold in Italy and King Lear Overture. Grande Valse di Bravura.
1837	*January-February:* L. introduces Chopin to George Sand. He gives chamber concerts in Paris with Urhan and Batta. *March 31:* Concert at Princess Belgiojoso's in which both L. and Thalberg play.	Soirées musicales (Rossini). Hexameron. Sketch of Dante Sonata. Transcriptions of Beethoven's Symphonies 5-7.

Year	*Life*	*Work*

1837 *May 1 to end of July:* L. and Marie d'Agoult visit George Sand's house at Nohant.
August 3: Concert in Lyon.
September: Both depart for Italy, visiting Bellaggio on Lake Como and also Milan, where L. sees Rossini and gives concerts on *December 3* and *10.*
December 25: L.'s second daughter, Cosima, born.

Esmeralda (Louise Bertin).

1838 L. and Marie d'Agoult settle in Milan, *February.* L. gives concerts there, *February 18* and *March 15.*
March: Venice, followed by eight concerts in Vienna, where he meets Clara Wieck. Return to Venice, *May,* and summer with the Countess in Lugano. Later, concerts at Milan, Florence, Bologna, etc. At Pisa L. sees Orcagna's frescoes and sketches Totentanz.

Mélodies hongroises (Schubert).
Paganini Studies (1st version).
12 Grandes Études.
Grand Galop chromatique.
Sposalizio, Il Penseroso, Petrarch Sonnets (-*39*).
Transcriptions: William Tell Overture, Nuits d'Été à Pausilippe (Donizetti), Soirées Italiennes (Mercadante).
Schubert: Gondelfahrer, 12 Songs, Schwanengesang, Lob der Tränen.
Sketch of Totentanz.

1839 *February-June:* L. and the Countess in Rome. First concert there. *March* at Prince Galitzin's; in *May* L. plays the organ at the church of S. Luigi dei Francesi. Friendship with Ingres.
May 9: L.'s son Daniel born in Rome.
Summer spent at Lucca and San Rossore.

Angiolin del biondo crin.
Valse mélancolique.
Venezia e Napoli. La Romanesca.
Transcriptions: Adelaide (Beethoven), Winterreise (Schubert).
Sonnambula Fantasy.

Year   *Life*   *Work*

1839 *October 3:* L. offers to make good
the amount still needed for the
Beethoven Memorial out of his
own pocket.
November: Relations with the
Countess become strained. She
goes back to Paris with the children, while L. gives six concerts in
Vienna. He then visits Hungary
for the first time since he was a
boy, playing at Pozsony (*December
21*) and Budapest (*December 27
and 29*). He is received with
great acclamation, and proposes
the foundation of a national conservatoire at Budapest.

1840 Plays in Budapest on *January 2, 4,
6, 8, 9, 11, 12.* On *January 4* he is
presented with the sword of
honour; on *January 11* he conducts
for the first time. Later he plays
at Györ and Pozsony, conducting
again in the latter town (*January
26*). In *February* he plays at Sopron and revisits Raiding; he also
plays in Vienna, Prague (*March 7*),
Dresden (*March 16*) and Leipzig
(*March 17, 24* and *30*); in the latter
city he plays Schumann's Carnaval—he had previously met Schumann in Dresden. In *April* he
plays in Metz and then Paris,
where he meets Wagner for the
first time. In *May* and *June* he
visits England, playing in London
on *May 8* and *11* and *June 9
and 22.*
Summer: Brussels and various
Rhineland towns. At Ems he
meets Meyerbeer.
Aug.-Nov: Second visit to England.
October: Six concerts in Hamburg.

Mazeppa (*138*).
Morceau de Salon.
Heroischer Marsch im
ungarischem Stil.
I Puritani; Introduction
and Polonaise.
Fantasy on Lucrezia
Borgia.
Transcriptions: Six
Sacred Songs of Beeth-
oven, Four Sacred
Songs of Schubert,
Songs of Mendelssohn,
Hussitenlied, Magyar
Dallok and Magyar
Rhapsodiák (-47).
Funeral essay on Paga-
nini.

Year	Life	Work
1840	He founds a pension fund for the orchestra there. Then third visit to England; performs before Queen Victoria at Windsor Castle; tours England and Scotland ; visit to Dublin.	
1841	Concerts at Brussels and Liège; then in Paris. *April 26:* Concert with Berlioz for the Beethoven monument at Bonn (meeting with Wagner). *May-July:* More concerts in England. *July:* Hamburg, Kiel, Copenhagen. Summer holiday with the Countess and children on the Isle of Nonnenwerth in the Rhine. Concert tour in Germany; meets Spohr at Cassel. *November:* First visit to Weimar; then concerts in Jena, Dresden, Leipzig, Halle, Berlin, etc.	Malediction (?) Fantasies on Norma, Don Giovanni, Robert the Devil. Paraphrase on God Save the Queen. Le Moine. Septet (Beethoven). Il m'aimait tant. Im Rhein. Lorelei. Galop de Bal. (?) Galop in A minor. Feuilles d'Album. Nonnenwerth. Freischütz Fantasy. Rheinweinlied. Students' Song from Faust. Reiterlied. Was ist des Deutschen Vaterland.
1842	Series of concerts in Berlin. Meeting with Cornelius. Becomes member of the Royal Prussian Academy of Arts and honorary Doctor of Königsberg University. Concerts in Mitau, Dorpat, Riga. *April 20:* First concert in Petersburg. *June 30:* Charity concert in Paris. L.'s Students' Song sung in German. *July 20:* Plays at a concert in honour of Grétry at Liège, then at Brussels.	Gottes ist der Orient (*90*, 12). Wir sind nicht Mumien (*90*, 3). Das düstre Meer umrauscht mich. Über allen Gipfeln ist Ruh (*75*). Mignons Lied. Der König in Thule. Comment, disaient-ils. Der du von dem Himmel bist.

Year	*Life*	*Work*
1842	*September:* Second summer holiday at Nonnenwerth.	Albumblatt in Waltz form.
	October: Concerts in Thuringia. Receives the freedom of Jena. Second visit to Weimar; is appointed Grand Ducal Director of Music Extraordinary. Tours Germany and Holland with Rubini; second meeting with Wagner in Berlin.	Valse a capriccio Lucia e Parisina. O quand je dors. Vergiftet sind meine Lieder. Petite Valse favorite. Le rossignol. Chanson bohémienne. Canzone Napolitana. Elegy on themes of Prince Louis Ferdinand. Mazurka d'un Amateur de Petersbourg. Transcription of some organ fugues of Bach.
1843	*January:* Concerts in Berlin, Breslau and other Silesian towns. Conducts The Magic Flute in Breslau, *February 1.*	Bist du. Trinkspruch. Ländler in A flat. Transcriptions: Oberon Overture, Autrefois (Wielhorsky), Tscherkessenmarsch from Russlan and Ludmila, Galop of Bulhakoff, Gaudeamus Paraphrase, Marcia funebre (Eroica Symphony).
	April-May: Concerts in Warsaw, Cracow, Petersburg, Moscow, etc. Meets Glinka and Dargomijsky. After a visit to Hamburg, spends third and last summer holiday with the Countess and children on Nonnenwerth; then whole family visits L.'s mother in Paris.	
	Autumn: Concerts in Germany.	
	End December: L. takes up his duties in Weimar.	Figaro Fantasy.
1844	*January 7:* L. conducts his first concert in Weimar, followed by several others in Weimar and other German towns (*January-March*). Correspondence with the Countess, in which a separation is decided on.	Soldiers' Song from Faust (*90*). Freudvoll und leidvoll. Die Vätergruft. Les Aquilons (*80*). Five Songs with French texts (*18*).

Year	Life	Work
1844	*April:* Concerts in Paris. After making arrangements for the education of his children, L. leaves for further concerts in the French provinces. Visit to Lamartine at the Château de Saint-Point. At Pau he again meets Caroline de Saint-Cricq, now Mme d'Artigaux. *October - December:* Concerts in Zürich, Spain and Portugal.	Hungarian Storm March. Transcriptions: Marche funèbre de Dom Sébastien, Chanson du Béarn, Faribolo pastour. Published: Buch der Lieder.
1845	*January-April:* Concerts in Spain and Portugal; then France and Switzerland. In Basle he meets Raff, who becomes his temporary secretary. *August:* Visit to Bonn for the Beethoven Festival; first performance of L.'s Beethoven Cantata, *August 13.* Concerts in Coblenz, etc., then taken ill at Cologne and goes to Baden-Baden. *Autumn:* Concerts in Freiburg and Eastern France.	Le forgeron. Feuille morte. La Terre, Les Flots (*80*). First Beethoven Cantata. Vorder Schlacht (*90*). Es rufet Gott (*90*). Nicht gezagt (*90*). 3 Songs from Wilhelm Tell. Grosse Konzertfantasie über spanische Weisen (*253*). Jeanne d'Arc au bûcher. Madrigal (*171a*). Sketch for 1st Ballade. Es rauschen die Winde. Wo weilt er? Hymne de l'enfant à son réveil (*19*). Ich möchte hingehn. Wer nie sein Brot (*297*). Schwebe, schwebe.
1846	L. in France, then Frankfurt, Weimar, Vienna (*March*), Brno, Prague. *Summer,* in and near Vienna; meets Nicolai, Balfe and Wallace, and plays at a charity concert organised by Johann Strauss, sen.	Pater noster (*21*). Ave Maria (*20*). Die lustige Legion Isten veled. Capriccio alla turca (*388*). Tarantella (Muette de Portici). Hungarian Rhapsody No. 1.

Year	*Life*	*Work*
1846	*October:* Visit to Hungary and Roumania; meets Erkel and investigates gipsy music. The Countess d'Agoult publishes her novel " Nélida."	Transcriptions: Leier und Schwert (Weber), Freischütz Overture, Jubel-overture, Six Songs of Schubert, Schubert's Marches, Müllerlieder (Schubert), Spanish Serenade (Festetics).
1847	Tour through the Danube countries and the Ukraine. *February:* Meets Princess Carolyne Sayn-Wittgenstein at Kiev. Visit to Woronince; then via Lwow, Czernowitz, Jassy, etc., to Constantinople, where he plays to the Sultan Abdul-Mejid. Ten concerts in Odessa, *July.* The Princess persuades L. to give up his career as a virtuoso; after a concert at Elisabethgrad, *September*, he spends the winter at Woronince with the Princess.	*(47-52)* Harmonies poétiques et religieuses. Paraphrase on the March for Abdul-Mejid Khan. Titan (final version). Hungarian Rhapsody No. 2. Konzertparaphrase *(457)*. Glanes de Woronince. Le juif errant. Ernani Paraphrase *(431a)*. Transcriptions: Songs of Dessauer, Zigeuner-polka (Conradi), Schwanengesang and March from Hunyadi László, Songs *(535-40)* for pf., Cujus animam, La Charité (Rossini). Published: O lieb, so lang du lieben kannst. Mazeppa Study.
1848	L. in Lwow, Cracow and Ratibor. *February:* Return to Weimar, where Conradi joins him. Princess Sayn-Wittgenstein decides to leave her husband and join L. She visits Weimar in order to ask the Grand Duchess Maria Paulovna to dispose her brother, Tsar	First sketch of Ce qu'on entend sur la montagne. Hungaria Cantata. Male chorus Mass (1st version). Kling leise, mein Lied. Oh pourquoi donc. Romance *(169)*.

Year	*Life*	*Work*
1848	Nicholas I, to grant her a divorce. L. visits Wagner at Dresden; Wagner returns the visit in *August*. L. conducts the Tannhäuser Overture, *November 12*, and insists that the whole opera be produced at Weimar.	Die Macht der Musik. Les Préludes. Weimars Toten. 1st Ballade. Trois Études de Concert. Hungarian Rhapsody No. 9 Arbeiterchor. Le vieux vagabond. Transcriptions: Er ist ge-kommen (Franz), Sep-tet (Hummel), Was-serfahrt und Jägers Abschied (Mendels-sohn), Salve Maria (Verdi), Tannhäuser Overture (Wagner), Einsam bin ich (Weber), Schlummer-lied (Weber), Wid-mung (Schumann).
1849	Princess Sayn-Wittgenstein's peti-tion for divorce is refused; she and L. take up residence together at the Altenburg. *February 16:* First performance of Tannhäuser at Weimar. L. also gives Schumann's Faust, Part II. *May:* Wagner, in flight from Dresden, concealed by L. at Weimar and hears rehearsal of Tannhäuser. *June:* Bülow visits L. for the first time. *August 28:* First performance of Tasso and the Goethe March under L. *September-December:* Visit to Heli-goland, Hamburg and Bad Eilsen.	Ce qu'on entend sur la montagne (1st version). Grosses Konzertsolo. Ernani Fantasy. Über allen Gipfeln (2nd version). Licht, mehr Licht. Revision of the Dante Sonata. Consolations. Both Piano Concertos and Totentanz completed (1st versions). Goethe March (Pf.). Funérailles. Canzonetta del Salvator Rosa. Tasso (1st version). Chorus of Angels from Faust.

Year	*Life*	*Work*
1849		Work on Sardanapale. Transcriptions: O du mein holder Abendstern (Tannhäuser), Wedding March and Dance of the Fairies (Midsummer Night's Dream), Les Patineurs, (-*50*). Remaining transcriptions from Le Prophète, An die ferne Geliebte (Pf.), Hallo from Tony (E. H. zu S.-C.-G.). Published: Goethe Songs of Beethoven, Songs of Franz.
1850	*January:* Raff comes to Weimar; L.'s mother also visits him there. *End February:* First performance of Ce qu'on entend (1st version). *April:* Joachim visits L., and in *October* becomes leader of the Weimar orchestra. *August 24:* First performance of Prometheus. *August 28:* First performance of Lohengrin under L. Other operas first performed at Weimar during the year; Il Conte Ory (Rossini), Abenteuer Karls II (Hoven), Das Corps der Rache (Saloman), La Favorita (Donizetti).	Valse Impromptu. Mazurka brillante. Ce qu'on entend (2nd version). Prometheus (*69* and *99*). Festchor zur Enthüllung des Herderdenkmals. Pater noster (*22*) Work on Harmonies poétiques et religieuses (*173*). Héroïde funèbre. Fantasy and Fugue on Ad Nos. Transcriptions: 6 Organ Preludes and Fugues (Bach), Bunte Reihe (David), La cloche sonne, Orchestration of Polonaise (Weber). Published: Hohe Liebe, Gestorben war ich (also for pf. as Liebesträume).

Year	Life	Work
1851	Works performed during the year: König Alfred (Raff), Harold in Italy (Berlioz), Polonaise (Weber-Liszt), Ce qu'on entend and Tasso (2nd versions), Overture The Bride of Messina (Schumann). *June:* Bülow comes to Weimar as L.'s pupil. *December:* His overture Julius Cæsar is performed under L. *December 14:* First performance of the Wanderer Fantasy (Schubert-Liszt) in Vienna.	Scherzo and March. Mazeppa (symphonic poem). Fantasy on The Ruins of Athens (Beethoven). Polonaises 1 and 2. Études d'exécution transcendante. Paganini Studies (2nd version). Wanderer Fantasy (pf. and orch.). Transcribed: Symphony No. 9 (Beethoven) for 2 pfs. Published: Hungarian Rhapsodies 1 and 2, De la Fondation Goethe à Weimar.
1852	*March 20:* First performance at Weimar of Berlioz' Benvenuto Cellini under L. Cornelius visits L. for a fortnight and returns in the autumn. Works performed include Wagner's Faust Overture, Byron's Manfred with Schumann's incidental music, L.'s Mass for male voices, Verdi's Ernani, and a week of Berlioz' music, at which the composer is present. *December:* Joachim resigns his post at Weimar.	(-53) Sonata. Soirées de Vienne. Hungarian Fantasia. Ab Irato. Transcriptions: Benediction et Serment (Cellini-Berlioz), Kirchliche Festouverture (Nicolai), Two Pieces from Tannhäuser and Lohengrin, Abendstern (vlc. and pf.). Published: 3 Caprices-Valses, Grand Duo Concertant (vl. and pf.).
1853	Agnes Street comes to Weimar; Laub is appointed leader of the orchestra. Works performed include The Flying Dutchman, Ce	An die Künstler (1st version). Festklänge. 2nd Ballade.

Year	*Life*	*Work*
1853	qu'on entend in yet another version, Fantasy on The Ruins of Athens, and Hungarian Fantasia (the two latter in Budapest). *June:* Brahms visits L. for 2-3 weeks. *July:* L. visits Wagner in Zürich. *October 3-5:* Music festival at Karlsruhe; L. gives the first performance of the Künstlerchor, then, together with Bülow, Cornelius, Joachim, Reményi and others, visits Wagner at Basle. They then all journey to Paris, where Wagner meets Cosima for the first time.	Domine salvum fac. Te Deum (*24*). Huldigungsmarsch. Orpheus. Transcriptions from King Alfred (Raff). Published: Ave Maria and Pater noster (*20, 21*), Hungarian Rhapsodies 3-15, Harmonies poétiques et religieuses (*173*).
1854	Works performed: Nibelungen (Dorn), Flight into Egypt (Berlioz), Orpheus, Les Préludes, Mazeppa, Tasso (final version), Alfonso and Estrella (Schubert), Festklänge, The Siberian Hunter (Rubinstein). *July:* Visit to Holland, Belgium and North Germany. Book on the Gipsies sketched.	Vom Fels zum Meer. Ce qu'on entend (final version). Berceuse (1st version). Hungaria. Faust Symphony (without final chorus). Benedetto sia'il giorno (unpublished song version). Transcriptions: Aus Lohengrin.
1855	*February:* Second Berlioz week at Weimar: B. conducts the first performance of L.'s E flat Concerto with L. as soloist, also Fantastic Symphony and Lélio. Other works performed in the year: Genoveva (Schumann), L.'s Ave Maria, Cornelius' Mass, Fantasy and Fugue on Ad Nos (in Merseburg), Psalm 13 (in Berlin). Tausig becomes L.'s pupil. Tsar	(-*56*) Dante Symphony. Gran Mass. New version of Prometheus. Beatitudes begun. Psalm 13. Fantasy and Fugue on BACH (1st version). Ein Fichtenbaum (I). Nimm einen Strahl der Sonne.

Year	*Life*	*Work*
1855	Alexander II deprives the Princess of her nationality and confiscates her fortune.	Was Liebe sei (II). Published: Années de Pèlerinage I, Articles on—Berlioz and his Harold Symphony, Robert Schumann, Clara Schumann, Robert Franz, Sobolewski's Vinvela, No Entr'acte Music!, The Rheingold, Marx and his book " Die Musik der 19 Jahrhunderts und ihre Pflege."
1856	*January:* L. conducts at the Mozart Centenary Festival in Vienna. *February-March:* Berlioz conducts the Corsair Overture and The Damnation of Faust at Weimar. *April:* Verdi's I Due Foscari at Weimar; L. receives the first numbers of the text of St. Elisabeth from Roquette. Bülow asks for Cosima's hand. *May:* First performance of BACH Fugue at Merseburg. *August 31:* First performance of Gran Mass at Gran (Esztergom). *September 8:* First performance of Hungaria at Budapest under L. Then visits to Vienna and Prague. *October:* L. and the Princess visit Wagner in Zürich; private performance of Act I of Walküre. After a short illness L. returns to Weimar via Munich, where he meets Kaulbach, whose painting, Die Hunnenschlacht, inspires L.'s symphonic poem.	Anfangs wollt' ich fast verzagen. Vereinslied (*90*). Completion of Dante Symphony. Concerto pathétique (from Grosses Konzertsolo). Festvorspiel (Preludio pomposo). New versions of: Angiolin, Am Rhein, Der du von dem Himmel bist, Lorelei, Kennst du das Land, Es war ein König in Thule, Wie singt die Lerche schön; also An die Künstler. Published: Tyrolean Melody (*233a*).

Year	Life	Work
1857	*January 7:* First performance of A major Concerto and final version of Ce qu'on entend at Weimar. *January 22:* First performance of Sonata by Bülow in Berlin. *February 26:* L. conducts several of his works at Leipzig; strong opposition in the Press. *May 31-June 2:* L. conducts at Lower Rhine Music Festival in Aachen; similar opposition led by Hiller. *August 18:* Bülow married to Cosima in L.'s presence. *September 4:* First performance of Festvorspiel in Weimar. *September 5:* First performance of Faust Symphony and Die Ideale, Weimar. *October 22:* Blandine married to Emile Ollivier. *November 7:* First performance of the Dante Symphony at Prague; it is a failure. *November 10:* First performance of Héroïde funèbre at Breslau. *December 29:* First performance of Hunnenschlacht under L., Weimar. The Grand Duke's interest in music begins to flag, and the intrigues of the Intendant Dingelstedt undermine L.'s work.	Hunnenschlacht completed. Final chorus of Faust Symphony. Die Ideale. Künstlerfestzug. Goethe March (2nd version). Weimars Volkslied. Es muss ein Wunderbares sein. Ich liebe dich. Ständchen (*90, 2*). Muttergottessträusslein. Work begun on St. Elisabeth.
1858	*March:* Visits to Prague and Budapest, where L. conducts some of his works. *May 27:* First performance of Festgesang zur Lehrerversamm-lung, Weimar. The critic Franz Brendel publishes his book, "Liszt als Symphoniker."	Hamlet. Lenore. Festgesang zur Lehrerversammlung. Revision of Jeanne d'Arc. Published: Années de Pèlerinage II.

Year	Life	Work

1858 *September 30:* First performance of
Sobolewski's Comala, under L.
December 15: First performance of
Cornelius' Barber of Bagdad
under L. This is received with a
hostile demonstration which L.
takes to be directed against himself; he tenders his resignation to
the Grand Duke.

1859 *January:* First disaffection between
Wagner and L. Bülow conducts
several of L.'s works in Berlin and
Prague.
April: First performance of the
Huldigungsmarsch at Weimar.
L. receives the Order of the Iron
Crown, and later is raised to the
Austrian nobility.
October: First performance of the
Beatitudes, Weimar.
November: First performance of
Vor Hundert Jahren.
December 13: Daniel Liszt dies in
Berlin. Princess Sayn-Wittgen-
stein falls out of favour with the
Weimar court, and L.'s position
becomes more and more unten-
able.

Vor hundert Jahren.
The Beatitudes (com-
pleted).
Festlied zu Schillers
Jubelfeier (*90*).
Venezia e Napoli (2nd
version).
Prelude, Weinen,
Klagen.
Psalms 23 and 137.
Morgenlied (*88*).
Te Deum (*27*).
Mephisto Waltz No. 1
(pf.).
Festmarsch nach Motiven
von E. H. zu S. C. G.
Rigoletto Fantasy.
Ernani Fantasy (revised).
Rienzi-Fantasiestück.
Miserere from Il Trova-
tore.
Orchestration of Schu-
bert's Marches.
Published: Lasst mich
ruhen, In Liebeslust.

1860 *March:* The protest against the
New Music, signed by Brahms,
Joachim, Grimm and Scholz, is
published.
May: The Princess leaves Weimar
and goes to Rome in the hope of

Ich scheide.
Les Morts.
Psalm 18.
Wieder möcht' ich dir
begegnen.
Responsorien.

Year	*Life*	*Work*
1860	obtaining her divorce from the Pope.	Die drei Zigeuner.

1860   obtaining her divorce from the
Pope.
August 15: First performance of
the Te Deum. L. made Officier
de la Légion d'Honneur.
November 8: First performance of
the Künstlerfestzug in Weimar.
L. makes his will, in the form of a
letter to the Princess.

Die drei Zigeuner.
Der traurige Mönch.
Pater noster from
   Christus (*3*, *7*).
An den heiligen Franzis-
   kus von Paula.
Two Episodes from
   Lenau's Faust.
Die stille Wasserrose.
Nonnenwerth (II).
Jugendglück.
Blume und Duft.
Wer nie sein Brot (II).
Mignons Lied (3rd ver-
   sion).
Transcriptions: Helges
   Treue (Draeseke), Pil-
   grims' Chorus from
   Tannhäuser (org.),
   Introduction and
   Fugue—ich hatte viel
   Bekümmernis (Bach),
   Spinning chorus from
   the Flying Dutchman,
   Danse des Sylphes
   (Berlioz), Festmarsch
   zur Schillerfeier
   (Meyerbeer); **Orches-**
   tration of Lorelei,
   Mignon, Schubert
   songs.
Published: 6 Chants
   Polonais (Chopin).

1861   *March 8:* First performance of
The Dance in the Village Inn (1st
Mephisto Waltz), Weimar.
May: Visit with the Bülows to
various South German towns.
June 25: First performance of
Psalm 18, Weimar.

Work on St. Elisabeth.
Löse, Himmel, meine
   Seele (Lassen).
Pastorale and Schnitter-
   chor from Prometheus.
Waltz from Gounod's
   Faust.

Year	Life	Work
1861	*August:* Conference held at Weimar for the foundation of the Allgemeine Deutsche Musikverein; Wagner, Cornelius and Bülow are present. *August 17:* L. leaves Weimar, and travels to Berlin and Paris, where he sees the Countess d'Agoult, and also meets Rossini, Gounod and Halévy; he plays at the Tuileries before Napoleon III and the Empress Eugénie. *October 21:* Arrival in Rome. The Princess hopes to obtain her divorce, and preparations for the wedding are made; but at the last moment the Pope revokes his sanction of the divorce. Both remain in Rome and find separate establishments.	Published: Für Männergesang, Variants to Festklänge, Geharnischte Lieder, An den Sonnenschein and Rotes Röslein (Schumann) (pf.).
1862	*September 11:* Blandine Ollivier dies. First performance of Cantico del Sol, Rome.	St. Elisabeth completed. Variations on Weinen, Klagen. Ave Maria (for the piano school of Lebert and Stark) (*182*). Hosannah for bass trombone and organ. Berceuse (2nd version). Alleluja and Ave Maria (Arcadelt). À la chapelle Sixtine. Cantico del Sol. Waldesrauschen and Gnomenreigen.
1863	*June:* L. enters the Oratory of the Madonna del Rosario, Monte Mario, Pope Pius IX visits him	Hirtenspiel and Heilige Drei Könige (Christus). Christus ist geboren.

Year	*Life*	*Work*

1863	there. First performance of Slavimo, Rome.	Slavimo.
		Rhapsodie espagnole.
		Legendes.
		Salve Polonia.
		Transcriptions:
		  Symphonies 1-4 and 8
		  (Beethoven).

1864	*March:* L. plays at a charity concert in Rome. *July:* L. visits Cardinal Hohenlohe at the Villa d'Este, and then goes to Castel Gandolfo, where he plays to the Pope. *August:* L. goes to the congress of the Allgemeine Deutsche Musikverein at Karlsruhe; there he meets Cosima, and visits Wagner on the Starnberger See. He plays him the Beatitudes. Then visit to Kaulbach in Munich. *September:* Visits to Weimar, Berlin and other German towns; then to Paris, where he sees the Countess d'Agoult. Return to Rome via the South of France. Princess Sayn-Wittgenstein's husband has died in *March*, but there is now no talk of marriage.	Vexilla regis. Urbi et Orbi. Ora pro nobis. La Notte. Transcription: Symphony No. 9 (Beethoven).
1865	*April 15:* First performance of the Totentanz, The Hague. *April 25:* L. receives the tonsure from Grand Almoner Prince Hohenlohe; he enters the Vatican, then spends *May* with Hohenlohe at the Villa d'Este. *May 25:* First performance of the Pater Noster from Christus, Dessau.	Inno del Papa (choral version). Missa choralis. Ave maris stella. Piano pieces 1 and 2 (*192*). Crux, Hymne des marins. Work on Christus; Introduction, Pastorale and Annunciation, Stabat mater speciosa, The

Year	Life	Work
1865	*July 30:* L. receives the next three minor orders. *August 15:* First performance of St. Elisabeth, Budapest, under L. *August 29:* First performance of the Two Legends, Budapest, played by L. *September:* Return to Rome.	Miracle, Entry into Jerusalem. Transcriptions: Weimars Volkslied (org.), Regina coeli (Lassus) (org.), Rákóczy March (orch.), Mazurka Fantasy (Bülow), Illustrations de l'Africaine (Meyer-beer). Published: Vom Fels zum Meer (pf.), À la Cha-pelle Sixtine, Ora pro nobis, Les Sabéennes (Gounod), Confutatis and Lacrymosa from Mozart's Requiem, Pil-grims' Chorus from Tannhäuser, Weihn-achtslied.
1866	*January 4:* First performance of Stabat mater speciosa, Rome. *February 6:* L.'s mother dies in Paris. *March-April:* Visit to Paris. Final break with Mme d'Agoult. L. hears Franck play at Ste. Clo-thilde. *May:* Return to Rome. Hohen-lohe is made a cardinal.	Le Triomphe funèbre du Tasse. Completion of Christus. Hungarian Coronation Mass. Male chorus added to Les Morts. Piano piece in A flat (*189*). Transcribed: Hymne à Ste. Cécile (Gounod). La Notte (vl. and pf.). Published: Löse, Himmel, meine Seele, Marche au supplice and Marche des pèlerins (Berlioz—new versions), Concerto pathétique, Danse des sylphes (Berlioz).

Year	Life	Work
1867	*June 8:* First performance of the Hungarian Coronation Mass, Budapest. *July 6:* First performance of the Christmas Oratorio from Christus, Rome. *August:* Visit to Meiningen and Weimar, where St. Elisabeth is performed; then visit to Wagner at Triebschen, and return to Rome, *October.*	Foundation of the Church (Christus). Hungarian Coronation Mass completed. Marche funèbre (for Maximilian of Mexico). Fantasy on Szep Ilonka (Mosonyi). Transcribed: Liebestod from Tristan, Benedictus and Offertorium from Hungarian Coronation Mass.
1868	*June 21:* L. plays before Pope Pius IX on the 20th anniversary of his coronation. *July-August:* L. in Grottamare with Don Antonio Solfanelli, who gives him serious religious instruction; then return to the Villa d'Este.	O filii (Christus). Ave maris stella (new versions). Mihi autem adhaerere. Requiem. La Marquise de Blocqueville. Technical Studies begun. Les Adieux (Romeo et Juliette—Gounod). Don Carlos Transcription (Verdi).
1869	L. returns to Weimar and begins his teaching activities in the Hofgärtnerei; Rubinstein visits him there. *May 15:* First (?) performance of Ave maris stella, Regensburg. *September 22:* L. attends the Rheingold performance in Munich which was put on against Wagner's wishes. *October 17:* First (?) performance of the Requiem, Lwow. *Winter:* L. at the Villa d'Este; meeting with Grieg in Rome.	2nd Beethoven Cantata. O salutaris hostia *(40).* Pater noster *(41).* Ave Maria *(38).* Ave Maria *(341).* Tantum ergo. Male chorus Mass (2nd version). Psalm 116. Gaudeamus igitur, Humoresque *(71).* Inno a Maria Vergine. Sketches for St. Stanislaus. Edition of Schubert's and Weber's sonatas.

Year	*Life*	*Work*
1870	*May 29:* L. conducts at Weimar in the Beethoven Centenary Festival (his 2nd Beethoven Cantata and Beethoven's 9th Symphony). *July:* Lenbach paints L., Munich. *August 25:* Wagner and Cosima are married at Lucerne. First performance of Gaudeamus igitur Humoresque, Jena.	BACH Fugue for organ. Hungarian March for the Coronation Festival. Ungarischer Geschwind- marsch. Mosonyis Grabgeleit. Szózat and Hymnus (Egressy and Erkel). Der Schwur am Rütli (Draeseke).
1871	*June 25:* L. appointed Royal Hun- garian Counsellor. From now till the end of his life, regular three- cornered journeys between Rome, Weimar and Budapest.	Ave verum corpus. Libera me. BACH Fugue for pf. Lied der Begeisterung. Die Fischerstochter. Transcriptions: Am stillen Herd (Meistersinger), Die Allmacht (Schu- bert). Published: Psalm 18.
1872	*September:* Wagner and Cosima visit L. in Weimar; their strained relations are eased, and L. pays his first visit to Bayreuth, *October.* *November:* Visit to Hungary, in- cluding Raiding, for the first time since 1848.	La perla. Sunt lacrimae rerum. Epithalam. Impromptu (Nocturne). J'ai perdu ma force et ma vie. Wartburglieder. Published: La Marseill- aise, Fantasy and Fugue in G minor (Bach), Ich weil' in tiefer Einsamkeit (Lassen). Frühlings- nacht (Schumann), Songs of Robert and Clara Schumann, Orch. version of Wil- liam Tell songs and Drei Zigeuner.

Year	*Life*	*Work*

1873 *March 2:* L. plays at a charity concert in Budapest.

May 29: First complete performance of Christus under L. at
Weimar.

August 2: L. attends the opening
celebration (?) of the Bayreuth
Festspielhaus.

September 7: L. plays before the
Grand Ducal pair in Weimar.

September 23: First performance of
the play Der Brautwillkomm auf
Wartburg with L.'s music (Wart-
burglieder).

November 8-11: Celebration of the
50th anniversary of L.'s artistic
career at Budapest; the event is
treated as a great national occasion.

Piano piece No. 3 (*192*).
Ave Maria (*504*).
Published: 5 Hungarian
Folk Songs, Ballade
(The Flying Dutch-
man).

1874 *January and February:* L. plays at
charity concerts in Vienna and
Sopron; Brahms also takes part in
the former.

First Elegy.
Ihr Glocken von Marling.
Und sprich.
Des toten Dichters Liebe.
Anima Christi sanctifica
me.
Die Glocken des Strass-
burger Munsters.
Die heilige Cäcilia.
Des erwachenden Kindes
Lobgesang (final ver-
sion).
Sankt Christoph.
Jeanne d'Arc (final orch.
version).
First scene of St. Stanis-
laus completed.
Weihnachtsbaum (*-76*).
Transcription: Dante
Sonnet (Bülow).

Year	*Life*	*Work*

1875 *March 10:* Concert given by Wagner in Budapest includes first performance of Die Glocken des Strassburger Munsters; L. also plays Beethoven's E flat concerto. *June 17:* Memorial ceremony for Frau von Moukhanoff at Weimar; various works of L. are played, including first performance of 1st Elegy. *July-August:* L. visits Bayreuth for the Festival rehearsals. L. becomes President of the newly founded Landesmusikakademie in Budapest.

Der Herr bewahret die Seelen seiner Heiligen. Der blinde Sänger. Ungarischer Sturmmarsch (Orch.). Benedictus from the Hungarian Coronation Mass (vln. and orch.). Published: An den heiligen Franziskus von Paula.

1876 *Mid-February:* L. in Budapest. *March 5:* Countess d'Agoult dies. *May:* Visit to King of Holland, then Weimar. *July 2:* First performance (?) of Hamlet, Sondershausen. *August:* Visit to Bayreuth Festival for performances of The Ring. *Winter:* Budapest, where he conducts a class of advanced piano playing at the Academy. At Bayreuth, meeting with Tchaikovsky.

Festpolonaise. Piano piece No. 4 (*192*). Weihnachtslied (*49*). Transcribed: Danse Macabre (Saint-Saëns). Published: Prometheus choruses (final version), Excelsior (pf. duet), Jeanne d'Arc (final version).

1877 *March:* First performance of Triomphe Funèbre du Tasse, New York. L. plays in a concert at Vienna for the fiftieth anniversary of Beethoven's death; Busoni hears him. Visits to Bayreuth and Weimar; at the latter Borodin comes to see him. Then Villa d'Este. *December:* L. gives first perform- of Saint-Saëns' Samson et Dalila,

Angelus. Aux Cyprès de la Villa d'Este I and II. Second Elegy. Les jeux d'eaux de la Villa d'Este. Sancta Dorothea. Resignazione (org.). Sei still. Dem Andenken Petöfis. Sursum corda.

Year	*Life*	*Work*
1877	Weimar; the composer and Fauré are present.	Transcribed: Agnus Dei from Verdi's Requiem (org.), Orch. of 2nd Overture to Cornelius' Barber of Bagdad. Published: Trois morceaux suisses (new ed.), Die Rose (Spohr), Valse d'Adèle (Zichy).
1878	Visits to Vienna, Weimar, Hanover, Paris, Bayreuth, Budapest (where he meets Albeniz) and the Villa d'Este.	Via crucis. Septem sacramenta. 12 alte deutsche geistliche Weisen. An Edlitam. Der Glückliche. Einst. Die tote Nachtigall (2nd version). Transcribed: Nibelungen and Faust music (Lassen), Der blinde Sänger (pf.), Introduction and Hungarian March (Széchenyi).
1879	*January:* Rome and Budapest. *February 8:* L.'s cousin Eduard von Liszt dies. *April:* Vienna; two of the Septem Sacramenta given first (?) performance; then Hanover, Frankfurt, Weimar. *June:* Wiesbaden. *July:* Sondershausen, for performances of his works. *July 10:* Septem Sacramenta complete given at Weimar. *August:* Bayreuth; then Villa d'Este till end of year. L. is made an honorary canon of St. Albano in Rome.	Rosario. Gebet (org.; *265*). Missa pro organo. Cantantibus organis. Ossa arida. O Roma nobilis. Revive Szegedin. Toccata. Sospiri. Carousel of Mme. Pelet-Narbonne. Sarabande and Chaconne from Handel's Almira. Was Liebe sei (3rd version). Go not, happy day. Transcribed: Tarantella (Dargomijsky). Published: Aïda Transcription.

Year	*Life*	*Work*
1880	*January:* Venice, then Budapest (meeting with Joachim). *March:* Vienna for a concert of his works; then Weimar. *September* till end of year, Villa d'Este, apart from meeting with Wagner at Siena, *September 16.*	San Francesco, Prelude. 2nd Mephisto Waltz. Verlassen. Des Tages laute Stimmen Schweigen. Angelus (str. quartet). Romance oubliée. In festo transfigurationis. Pro Papa. Variation on Chopsticks. Transcribed: Polonaise from Eugene Onegin (Tchaikovsky), Die sieben Todssünden (Goldschmidt). Published: Walhall (Rheingold).
1881	*January:* Via Florence to Budapest; meeting with Brahms there, *February.* *March 9:* First performance of 2nd Mephisto Waltz, Budapest. Visit to Sopron and Raiding; then Berlin, where concerts of his works are given, and he meets Bülow again. *May:* Visits to Cologne and Antwerp. *June:* Bülow visits L. at Weimar and meets Cosima again. *July:* L. falls down the stairs at Weimar—this accident has lasting consequences. *September-October:* Bayreuth, then Rome. *October 22:* L.'s 70th birthday is celebrated in Rome with concerts of his works. L. meets Tchaikovsky again. *December 25:* First performance of Weihnachtslied (*49*).	Wiegenlied (*198*). Nuages gris. Ave Maria (*341*). Valse oubliée No. 1. Ungarns Gott. Von der Wiege bis zur Grabe. Cantico del sol (pf.). O Meer im Abendstrahl. Psalm 129. Transcribed: Danses galiciennes (Zarembski), Cantico del sol (org.), Provençalisches Minnelied (Schumann), O wenn es doch immer so bliebe (Rubinstein).

Year	*Life*	*Work*

1882 *January:* Venice, Vienna.
February: Budapest.
April: First performance of Angelus for string quartet, Weimar.
May: Visit to Brussels and Antwerp (performances).
July: Visits to Freiburg, Baden-Baden, Zürich, for performances.
August: Bayreuth for the first 5 Parsifal performances, then Weimar till *mid-November*; Rubinstein and Nikisch visit him there.
November: Joins Wagner and Cosima in Venice.

Csárdás macabre.
Valse oubliée No. 2.
Hungarian Rhapsody No. 16.
La lugubre gondola I and II.
Réminiscences de Boccanegra (Verdi).
Transcribed: Über allen Zaubern Liebe (Lassen), Tannhäuser-Lieder (Lessmann), Feierlicher Marsch zum Gral (Parsifal).
Published: 12 Kirchenchorgesänge.

1883 *January 13:* Leaves Venice for Budapest, shortly before Wagner's death (*February 13*).
March: Vienna, Weimar. L. conducts in a memorial concert for Wagner in Weimar on *May 22*, and remains there till the end of the year.

R.W.-Venezia.
Am Grabe Richard Wagners.
Schlaflos.
Requiem for organ.
Nun danket alle Gott.
Zur Trauung.
Ungarisches Königslied.
Hungarian Rhapsody No. 17.
Bülow-Marsch.
Mephisto Polka.
Mephisto Waltz No. 3.
Valse oubliée No. 3.
Cadenzas to Nos. 6 and 9 of Soirées de Vienne.
Transcribed: Two Songs (Korbay).
Published: Années de Pèlerinage III, 3 Petrarch Sonnets for baritone (new version), Salve Polonia, Kavalleriegeschwindmarsch.

Year	Life	Work
1884	*February:* Budapest and various Hungarian towns. *April:* Vienna and Weimar. *May:* Congress of Allgemeine Deutsche Musikverein at Weimar; L. conducts for the last time in Weimar. *July:* Bayreuth (Parsifal). *August:* Munich, Weimar. *October:* Vienna, Budapest. *December:* Rome.	Le Crucifix. Qui seminant in lacrimis. Introitus (org.). 2 Csárdás. Mariengarten. Und wir dachten der Toten. Sanct Caecilia. In domum Domini. Published: Recueillement, Der Asra (Rubinstein).
1885	*January:* Rome, Florence, then Budapest till *Easter*. *April:* Vienna, Weimar. Visits to Mannheim, Karlsruhe, Strasbourg, Antwerp, Aachen for performances, etc. Then Weimar till *September*; afterwards Munich, Innsbruck, Rome.	Historische ungarische Bildnisse. Abschied (*251*). Pax vobiscum. 4th Mephisto Waltz. Trauervorspiel und Trauermarsch. Qui Mariam absolvisti. Gruss (*94*). Salve regina (*66*). Valse oubliée No. 4. Bagatelle sans Tonalité. Hungarian Rhapsodies 18 and 19. En rêve. Unstern. Transcribed: Tarantella (Cui). Published: Pusztawehmut, Pilgrims' Chorus from Tannhäuser (2nd version).
1886	*January:* L. leaves Rome for the last time; thence Florence, Venice, Budapest. Farewell concerts at the last-named. *March:* Concerts of L.'s works at Liège and Paris.	Orch. of Die Vätergruft. Published: Hungarian Rhapsodies 17 and 19.

Year	*Life*	*Work*

1886 *April:* Visit to London (the first
since 1841). Performance of St.
Elisabeth under Mackenzie, and
several other Liszt concerts. L. is
received by Queen Victoria.

April-May: Antwerp and Paris for
further performances. L. catches
cold and becomes very weak.

Middle of May: Weimar.

June: Sondershausen, then to consult doctor in Halle; dropsy is
diagnosed.

July: Visit to Bayreuth for the
wedding of his granddaughter,
Daniela von Bülow; then to Col-
pach in Luxemburg. On *July 19*
L. plays for the last time, at a
concert in Luxemburg.

July 21: Arrival in Bayreuth. He
hears Parsifal on *23rd* and Tristan
on *25th*. His illness develops into
pneumonia.

July 31: Death of Liszt.

1887 *March 8:* Death of Princess Sayn-
Wittgenstein.

CATALOGUE OF WORKS

This catalogue attempts to include all Liszt's works, except for some short pieces written for private albums and not so far published; the numbering corresponds with that of the catalogue prepared by the author for the Fifth Edition of " Grove's Dictionary of Music " (by kind permission of the Editor), and readers are referred to the latter for further information about the works. A good deal of useful information may also be found in the catalogue in the second volume of Peter Raabe's " Franz Liszt " (Stuttgart, 1931). Each work is followed by the date of composition, or, if this is unknown, of publication; in addition, a Roman followed by an Arabic numeral (thus II,2) indicates the section and volume of the Breitkopf Collected Edition in which the work is published. Works published by the Liszt Society (Schott & Co., London) are indicated by L.S. I, II, etc.

A. ORIGINAL WORKS

1. OPERA

1. Don Sanche, ou Le Château d'Amour. Opéra en un acte. 1824-5.

2. SACRED CHORAL WORKS

2. The Legend of St. Elisabeth. Oratorio on words of Otto Roquette. 1857-62.

3. Christus. Oratorio on texts from the Holy Scripture and the Catholic Liturgy. 1855-67.
 I. Christmas Oratorio.
   1. Introduction. 2. Pastorale and Annunciation. 3. Stabat mater speciosa. 4. Shepherds' song at the manger. 5. The Three Holy Kings (March).
 II. After Epiphany.
   6. The Beatitudes. 7. Pater noster. 8. The Foundation of the Church. 9. The Miracle. 10. The Entry into Jerusalem.

III. Passion and Resurrection.
      11. Tristis est anima mea. 12. Stabat mater
      dolorosa. 13. O filii et filiae. 14. Resurrexit.

4. Cantico del sol di Francesco d'Assisi. (Bar. solo, male chorus,
    orch. and organ. 1862; rev. 1880-1.   V, 5.

5. Die heilige Cäcilia. Legende. M-S solo, chorus and orch.
    1874.

6. Die Glocken des Strassburger Münsters (Longfellow). M-S,
    Bar. soli, chorus and orch. 1874.
      1. Vorspiel; Excelsior.
      2. Die Glocken.

7. Cantantibus organis. Antifonia per la festa di Sta. Cecilia.
    Solo, chorus and orch. 1879.   V, 5.

8. Missa quattuor vocum ad aequales concinente organo. Male
    chorus and organ. 1st version 1848; 2nd version 1869. V, 3

9. Missa solennis zur Einweihung der Basilika in Gran (Gran
    Mass). SATB soli, chorus and orch. 1855; rev. 1857-8.

10. Missa choralis, organo concinente. Mixed chorus and organ.
    1865.   V, 3

11. Hungarian Coronation Mass. SATB soli, chorus and orch.
    1867 (Graduale 1869). Cf. *15a.*

12. Requiem. TTBB soli, male chorus, organ and brass. 1867-8;
    " Libera me." 1871.   V, 3

13. Psalm 13. " Lord, how long ? " T solo, chorus and orch.
    1855, rev. 1859.

14. Psalm 18. " Coeli enarrant." Male chorus and orch. 1860.

15. Psalm 23. " The Lord is my shepherd." T or S solo, harp,
    organ; also with male chorus ad lib. 1859; rev. 1862.

15a. Psalm 116. " Laudate Dominum." Male chorus and pf. 1869;
    cf. *11.*

16. Psalm 129. " De Profundis." Bar. solo, male chorus and
    organ; or B (or A) and pf. 1881.

17. Psalm 137. " By the waters of Babylon." S solo, women's
    chorus, solo violin, harp and organ. 1859; rev. 1862.

18. Five choruses with French texts. (In the 1840s.)
      1. Qui m'a donné. 2. L'Éternel est son nom (Racine).
      3. Chantons, chantons l'auteur. 4. (A major, without text.)
      5. Combien j'ai douce souvenance (Chateaubriand).

19. Hymne de l'enfant à son réveil (Lamartine). Women's
    chorus, harm. and harp. *C.* 1845; rev. 1862 and 1874. V, 5

20. Ave Maria I. 1st version, chorus and organ, 1846.   V, 6
                      2nd version, " Quattuor vocum concinente
                      organo," *c.* 1852.   V, 6

21. Pater noster II. 1. Male chorus unacc., 1846.
2. Pater noster quattuor vocum adaequales concinente organo secundum rituale SS. ecclesiae Romanae, *c.* 1848. V, 6

22. Pater noster IV. Mixed chorus and organ. 1850.

23. Domine salvum fac regem. T solo, male chorus and organ. 1853. V, 5

24. Te Deum II. Hymnus SS Ambrosii et Augustini. Male chorus and organ. 1853 (?). V, 7

25. Die Seligkeiten. Bar. solo, mixed chorus and organ. 1855-9; cf. *3,* 6. V, 6

26. Festgesang zur Eröffnung der zehnten allgemeinen deutschen Lehrerversammlung (Hoffmann von Fallersleben). Male chorus and organ. 1858. V, 6

27. Te Deum I. Mixed chorus, organ, brass and timps. 1859 (?). V, 7

28. An den heiligen Franziskus von Paula. Gebet. Male voices, organ, brass and timps. 1860 (at latest); rev. *c.* 1874. Cf. *175,* 2. V, 5

29. Pater noster I. Mixed chorus and organ. 1860 (at latest). Cf. *3,* 7. V, 6

30. Responses and Antiphons. Mixed chorus and organ. 1860. V, 7

31. Christus ist geboren I (Landmesser). Prob. 1863. V, 6
1. Mixed chorus and organ. 2. Male chorus and organ.

32. Christus ist geboren II (Landmesser). 1863 (?). V, 6
1. Mixed chorus and organ. 2. Male chorus unacc. (organ postlude). 3. SSA unacc.

33. Slavimo Slavno Slaveni! Male chorus and organ. 1863. V, 6

34. Ave maris stella. 1. Mixed chorus and organ. 1865 or 6. V, 6
2. Male chorus and organ. 1868. V, 6

35. Crux! Hymne des marins, avec Antienne approbative de N.T.S.P. Pie IX. 1865. V, 6
1. Male voices unacc. 2. Women's voices and pf.

36. Dall' alma Roma. 2-part chorus and organ. After 1867, from *3,* 8.

37. Mihi autem adhaerere. Male chorus and organ. 1868. V, 6

38. Ave Maria II. Mixed chorus and organ. 1869. V, 6

39. Inno a Maria Vergine. Mixed chorus, harp, organ or pf., 4 hands, and harm. 1869. V, 5

40. O salutaris hostia I. Female chorus and organ. Prob. 1869. V, 6

41. Pater noster III. 1. Mixed chorus and organ (F ma.). 1869.
                                   V, 6
          2. Male chorus and organ or harm. or pf.
          1869.   V, 6

42. Tantum ergo. 1. Male chorus and organ. 2. Female chorus
    and organ. 1869.   V, 6

43. O salutaris hostia II. Mixed chorus and organ. *C.* 1870 (?).
                                   V, 6

44. Ave verum corpus. Mixed chorus and organ (ad lib.). 1871.
                                   V, 6

45. Libera me. Male chorus and organ. 1871. Cf. *12.*

46. Anima Christi sanctifica me. Male chorus and organ. (2 ver-
    sions.) 1874.   V, 6

47. Sankt Christoph. Legend for Bar. solo, women's chorus, pf.,
    harm. and harp. 1874.

48. Der Herr bewahret die Seelen seiner Heiligen. Festgesang
    zur Enthüllung des Carl-August Denkmals in Weimar am
    3 September 1875. Mixed chorus, organ and wind. 1875.
                                     V, 6

49. Weihnachtslied (O heilige Nacht). T solo, women's chorus
    and organ or harm. After 1876; from *186,* 2.   V, 6

50. Chorales:
    1. Es segne uns Gott. 2. Gott sei uns gnädig. 3. Nun ruhen
    alle Wälder. 4. O Haupt voll Blut. 5. O Lamm Gottes.
    6. Was Gott tut. 7. Wer nur den lieben Gott. 8. Vexilla
    regis. 9. Crux benedicta. 10. O Traurigkeit. 11. Nun
    danket alle Gott. 12. Jesu Christe. 1878-9. 1 and 2 for
    mixed chorus and organ; remainder unacc. chorus. Cf. *51,*
    53 and *61.*   V, 7 (1-7 only)

51. Gott sei uns gnädig und barmherzig. Kirchensegen. Mixed
    chorus and organ. 1878; cf. *50,* 2.

52. Septem Sacramenta. Responsoria cum organo vel harmonio
    concinente.
    1. Baptisma. 2. Confirmatio. 3. Eucharistia. 4. Poeni-
    tentia. 5. Extrema Unctio. 6. Ordo. 7. Matrimonium.
    M-S, Bar. soli, mixed chorus and organ. 1878.   V, 7

53. Via Crucis. Les 14 Stations de la Croix pour Choeur et Soli,
    avec accompagnement d'orgue (ou pianoforte). 1878-9.
                                       V, 7

54. O Roma nobilis. Mixed chorus and organ or solo voice and
    organ. 1879.   V, 7

55. Ossa arida. Male chorus and organ, 4 hands, or pf., 4 hands.
    1879.   V, 6

56. Rosario. 1. Mysteria gaudiosa. 2. Mysteria dolorosa. 3. Mysteria gloriosa. 4. Pater noster. 1-3, mixed chorus and organ; 4, Bar. solo, male chorus and organ. 1879.　V, 7

57. In domum Domini ibimus. Mixed chorus, organ, brass and drums. (In L.'s last years.)

58. O sacrum convivium. A solo, women's chorus and organ. (In L.'s last years.)　V, 6

59. Pro Papa. 1. Dominus conservet eum. Mixed chorus and organ. *C.* 1880.　V, 6

　　2. Tu es Petrus. Male chorus and organ. (No conn. with *3*, 8.)

60. Zur Trauung. Geistliche Vermählungsmusik. (Ave Maria III.) Women's chorus and organ. 1883, from *161*, 1. V, 6

61. Nun danket alle Gott. Mixed or male chorus, organ, brass and drums. 1883.　V, 7

62. Mariengarten. (Quasi cedrus.) SSAT and organ. 1884. V, 6

63. Qui seminant in lacrimis. Mixed chorus and organ. 1884.
　　　　　　　　　　　　　　　　　　　　V, 6

64. Pax vobiscum! Male chorus and organ. 1885.　　V, 6

65. Qui Mariam absolvisti. Bar. solo, mixed chorus and organ. 1885.　　　　　　　　　　　　　　　　　V, 6

66. Salve Regina. Mixed chorus unacc. (No conn. with *669*, 1.) 1885.　　　　　　　　　　　　　　　　　V, 6

3. Secular Choral Works

67. Festkantate zur Enthüllung des Beethoven-Denkmals in Bonn (O. L. B. Wolff). SSTTBB soli, chorus and orch. 1845.

68. Zur Säkularfeier Beethovens (2nd Beethoven Cantata) (Adolf Stern and Gregorovius). SATB soli, chorus and orch. 1869-70.

69. Chöre zu Herders Entfesseltem Prometheus. SATTBB soli, chorus and orch. 1850; rev. 1855.

70. An die Künstler (Schiller). TTBB soli, male chorus and orch. 1st and 2nd versions, 1853; final version 1856.

71. Gaudeamus igitur. Humoreske. Soli (ad lib.), mixed or male chorus and orch. 1869.

72. Four-part male choruses (for the benefit of the Mozart-Stiftung).
　　　1. Rheinweinlied (Herwegh). 2. Students' song from Goethe's Faust. 3 and 4. Reiterlied (Herwegh), 1st and 2nd versions. 1 and 3 with pf., 2 and 4 unacc. 1841.

73. Es war einmal ein König (Goethe's Faust). B solo, male chorus and pf.

74. Das deutsche Vaterland (Arndt). 4-part male chorus. 2 versions. 1841.

75. Über allen Gipfeln ist Ruh (Goethe). Male chorus. 1st version, unacc., 1842; 2nd version, with 2 horns, 1849. Cf. *306.*

76. Das düstre Meer umrauscht mich. Male chorus and pf. 1842.

77. Die lustige Legion. (Buchheim.) Male chorus and pf. 1846.

78. Trinkspruch. Male chorus and pf. 1843.

79. Titan (Schober). Bar. solo, male chorus and pf. 1842, 1845, 1847.

80. Les Quatre Élémens. (Autran.) 1. La Terre. 2. Les Aquilons. 3. Les Flots. 4. Les Astres. Male chorus and pf. 1844-5. Cf. *97.*

81. Le Forgeron (Lamennais). Male chorus and pf. 1845.

82. Arbeiterchor. B solo, male quartet and chorus, pf. 1848. Cf. *100.*

83. Ungaria-Kantate (Schober). Bar. solo, mixed chorus and pf. 1848.

84. Licht, mehr Licht (Schober (?)). Male chorus and brass. 1849.

85. Chorus of Angels from Goethe's Faust. Mixed chorus and harp or pf. 1849.

86. Festchor zur Enthüllung des Herder-Denkmals in Weimar (Schöll). Male chorus and pf. 1850.

87. Weimars Volkslied (Cornelius). 1857, from *357.*
1. Male chorus and wind. 2. Male chorus and pf. 3. 4-part chorus. 4. Male chorus and organ. 5. Populaire. 6. 3-part chorus.

88. Morgenlied (Fallersleben). Women's chorus unacc. 1859.

89. Mit klingendem Spiel. Children's voices. *C.* 1859.

90. Für Männergesang.
1. Vereinslied (Fallersleben). 1856. 2. Ständchen (Rückert), with T solo. 1857. 3. Wir sind nicht Mumien (Fallersleben). 1842. 4-6. Geharnischte Lieder: Vor der Schlacht, Nicht gezagt, Es rufet Gott. (Meyer and Götze). 1845, with pf. Rev. *C.* 1860, unacc. 7. Soldiers' song from Goethe's Faust. (Trumpets and timps. ad lib.) 1844. 8. Die alten Sagen kunden (with solo quartet). *C.* 1845. 9. Saatengrün (Uhland). *C.* 1845. 10. Der Gang um Mitternacht (Herwegh) with T solo. *C.* 1845. 11. Festlied zu Schillers Jubelfeier (Dingelstedt), with Bar. solo. 1859. 12. Gottes ist der Orient (Goethe). 1842.

91. Das Lied der Begeisterung. A lelkesedes dala. 1871.

92. Carl August weilt mit uns. Festgesang zur Enthüllung des Carl-August-Denkmals in Weimar am 3 September 1875. Male chorus, brass, drums and organ. 1875. Cf. *48.*

93. Ungarisches Königslied. Magyar Király-dal. (Ábrányi.) 1883.
    1. Male chorus unacc. 2. Mixed chorus unacc. 3. Male chorus and pf. 4. Mixed chorus and pf. 5. Male or mixed chorus and orch.

94. Gruss. Male chorus. 1885 (?).

4. ORCHESTRAL WORKS

SYMPHONIC POEMS

95. Ce qu'on entend sur la montagne (Bergsymphonie) after Victor Hugo. 3 versions: 1. 1848-9. 2. 1850. 3. 1854.   I, 1

96. Tasso, Lamento e Trionfo, after Byron. 4 versions; 1849-54.   I, 1

97. Les Préludes, after Lamartine. 1848, as intro. to *80*; rev. before 1854. Cf. *142, 304.*   I, 1

98. Orpheus. 1853-4.   I, 2

99. Prometheus. 1850, as overture to *69*; rev. 1855. Cf. *121.*   I, 3

100. Mazeppa, after Victor Hugo. 1851; rev. *c.* 1854. From *139*, 4.   I, 3

101. Festklänge. 1853.   I, 4

102. Heroïde Funèbre. 1848-50, from *690*; rev. *c.* 1854.   I, 4

103. Hungaria. 1854, from *231*.   I, 5

104. Hamlet. 1858.   I, 5

105. Hunnenschlacht, after Kaulbach. 1856-7.   I, 6

106. Die Ideale, after Schiller. 1857; cf. *70.*   I, 6

107. From the Cradle to the Grave. 1881-2. Cf. *198.*   I, 10

OTHER ORCHESTRAL WORKS

108. A Faust Symphony in three character pictures, after Goethe. 1854; final chorus added 1857.   I, 8 and 9
    1. Faust. 2. Gretchen. 3. Mephistopheles, and final chorus.

109. A Symphony to Dante's Divina Commedia. 1. Inferno. 2. Purgatorio and Magnificat. 1855-6, with two endings.   I, 7

110. Two Episodes from Lenau's Faust. 1. Der nächtliche Zug. 2. Der Tanz in der Dorfschenke (1st Mephisto Waltz). 1860. Cf. *514.*   I, 10

111. Second Mephisto Waltz. 1880-1.   I, 10

112. Trois Odes Funèbres.
    1. Les Morts (Lamennais). Oration for full orchestra with
    male chorus. 1860.   I, 12
    2. La Notte, after Michelangelo. 1863-4, from *161*, 2.   I, 12
    3. Le Triomphe funèbre du Tasse, Epilogue to Tasso, *96*.
    1866.   I, 2
113. Salve Polonia. Interlude from the oratorio St. Stanislas, *688*.
    1863.
114. Künstlerfestzug zur Schillerfeier. 1857, on themes from *70* and
    106.   I, 11
115. Festmarsch zur Goethejubiläumsfeier. 1849; rev. 1857. Cf.
    227.   I, 11
116. Festmarsch nach Motiven von E.H.z.S.-C.-G. (Ernst Herzog
    zu Sachsen-Coburg-Gotha—from his opera Diana von
    Solange). 1859.
117. Rákóczy March, symphonic arrangement. 1865; cf. *244*, 15
    (sketched earlier).
118. Ungarischer Marsch zur Krönungsfeier in Ofen-Pest am 8
    Juni 1867. 1870(!)   I, 12
119. Ungarischer Sturmmarsch. 1875, from *232*.   I, 12

5. PIANOFORTE AND ORCHESTRA

120. Grande Fantaisie Symphonique on themes from Berlioz' Lélio.
    1834.
121. Malediction, with string orchestra. Sketched *c.* 1830 (?), rev.
    c. 1840(?). Cf. *99*, *108*.   I, 13
122. Fantasia on themes from Beethoven's Ruins of Athens. 1848-52.
123. Fantasia on Hungarian Folk Themes. *C.* 1852, from *244*, 14.
124. Concerto No. 1 in E flat. Sketched *c.* 1830; completed
    1849-56.   I, 13
125. Concerto No. 2 in A major. 1839; rev. 1849-61.   I, 13
126. Totentanz. Paraphrase on the Dies Irae. Planned 1838;
    (i) 1849; (ii) rev. 1853 and 1859.   I, 13

6. CHAMBER MUSIC, ETC.

127. Duo (Sonata), violin and piano. (In the 1830s.) On Chopin's
    Mazurka in C sharp minor, Op. 6, No. 2.
128. Grand Duo Concertant sur la Romance de M. Lafont, " Le
    Marin." *C.* 1835; rev. *c.* 1849. Violin and piano.
129. Epithalam zu E. Reményis Vermählungsfeier, violin and
    piano. 1872.
130. Elegie. 1. Cello, pf., harp and harm. 2. Cello and pf. 3. Violin
    and pf. 1874. Cf. *196*.

131. Second Elegy. Piano and violin or cello. 1877. Cf. *197.*

132. Romance Oubliée. Piano and viola or violin or cello. 1880; cf. *169, 527.*

133. Die Wiege (The Cradle). 4 violins. 1881. Cf. *107, 198.*

134. La Lugubre Gondola. Piano and violin or cello. 1882. Cf. *200,* 2.

135. Am Grabe Richard Wagners. String quartet and harp. 1883; cf. *6,* 1 and *202.*   L.S. II

7. PIANOFORTE, TWO HANDS

STUDIES

136. Étude en 48 exercices dans tous les tons majeurs et mineurs. 1826; only 12 were written. Cf. *137-9.*   II, 1

137. 24 Grandes Études. *C.* 1838; from *136.* Only 12 were written.
                                                                           II, 1

138. Mazeppa. *C.* 1840(?) from *137,* 4.   II, 1

139. Études d'exécution transcendante. 1851; from *137-8.* Cf. *100.*
                                                                           II, 2

    1. Preludio. 2. A minor. 3. Paysage. 4. Mazeppa. 5. Feux follets. 6. Vision. 7. Eroica. 8. Wilde Jagd. 9. Ricordanza. 10. F minor. 11. Harmonies du soir. 12. Chasse-neige.

140. Études d'exécution transcendante d'après Paganini. 1838; cf. *141.*   II, 3

    1. G minor. 2. E flat major. 3. La Campanella. 4. E major. 5. La Chasse. 6. Theme and Variations.

141. Grandes Études de Paganini. 1851, from *140.*   II, 3

142. Morceau de Salon, Étude de Perfectionnement. 1840; cf. *97, 143.*   II, 3

143. Ab Irato. Étude de Perfectionnement de la Méthode des méthodes. 1852, from *142.*   II, 3

144. 3 Études de Concert. 1. A flat major. 2. F minor. 3. D flat major. *C.* 1848. (Sometimes known as Il lamento, La leggierezza, Un sospiro.) *C.* 1848.   II, 3

145. 2 Concert Studies. 1. Waldesrauschen. 2. Gnomenreigen. 1862-3.   II, 3

146. Technical Studies (12 books). 1868 to *c.* 1880.

VARIOUS ORIGINAL WORKS

147. Variation on a waltz by Diabelli. 1822.   II, 7

148. Huit Variations. *C.* 1824.   II, 7

149. Sept Variations brillantes sur un thème de G. Rossini. *C.* 1824.

150. Impromptu brillant sur des thèmes de Rossini et Spontini. 1824.

151. Allegro di bravura. 1824.   II, 7
152. Rondo di bravura. 1824.   II, 7
153. Scherzo in G minor. 1827.   II, 9
154. Harmonies poétiques et religieuses. 1834. Cf. *173*, 4.
                                        L.S. II; II, 5
155. Apparitions. 1. Senza lentezza quasi Allegretto. 2. Viva-
mente. 3. Molto agitato ed appassionato. 1834; 3 on a
waltz of Schubert; cf. *427*, 4.   L.S. II; II, 5
156. Album d'un Voyageur. 1835-6; cf. *157,160*. (Lyon, L.S. II) II, 4
    I. Impressions et poésies.
          1. Lyon. 2a. Le Lac de Wallenstadt. 2b. Au bord
          d'une source. 3. Les cloches de G. . . . 4. Vallée
          d'Obermann. 5. La Chapelle de Guillaume Tell.
          6. Psaume.
    II. Fleurs mélodiques des Alpes.
          7a. Allegro. 7b. Lento. 7c. Allegro pastorale. 8a.
          Andante con sentimento. 8b. Andante molto
          espressivo. 8c. Allegro moderato. 9a. Allegretto.
          9b. Allegretto. 9c. Andantino con molto sentimento.
    III. Paraphrases.
          10. Improvisata sur le Ranz de Vaches de Ferd.
          Huber. 11. Un soir dans les montagnes. 12. Ron-
          deau sur le Ranz de Chèvres de Ferd. Huber.
157. Fantaisie romantique sur deux mélodies suisses. 1835; cf. *156*,
    7b.   II, 5
158. Tre Sonetti del Petrarca, 1st version. *C.* 1839(?), from *270*;
    cf. *161*, 4-6.   II, 5
159. Venezia e Napoli, 1st version. 1. Lento. 2. Allegro. 3.
    Andante placido. 4. Tarantelles napolitaines. *C.* 1840;
    cf. *96, 162*.   II, 5
160. Années de Pèlerinage. Première Année; Suisse. 7, 1836;
    the rest, 1848-54, from *156*.   II, 6
        1. Chapelle de Guillaume Tell. 2. Au Lac de Wallen-
        stadt. 3. Pastorale. 4. Au bord d'une source. 5. Orage.
        6. Vallée d'Obermann. 7. Eglogue. 8. Le mal du pays.
        9. Les cloches de Genève.
161. Années de Pèlerinage. Deuxième Année; Italie.   II, 6
        1. Sposalizio. 2. Il Penseroso. 3. Canzonetta del Salvator
        Rosa. 4. Sonetto 47 del Petrarca. 5. Sonetto 104 del
        Petrarca. 6. Sonetto 123 del Petrarca. 7. Après une lecture
        du Dante, Fantasia quasi Sonata.
            1, 2, 1838-9; 3, 1849; 4-6, from *158*, after 1846; 7 sketched
        1837, rev. 1849.

162. Venezia e Napoli. Supplement aux Années de Pèlerinage 2de
volume. 1859, from *159*, 3 and 4.   II, 6
   1. Gondoliera. 2. Canzone. 3. Tarantella.

163. Années de Pèlerinage. Troisième Année.   II, 6
   1. Angelus! Prière aux anges gardiens. 2. Aux cyprès de la
Villa d'Este, Thrénodie I. 3. Aux cyprès de la Villa d'Este,
Thrènodie II. 4. Les jeux d'eaux à la Villa d'Este. 5. Sunt
lacrymae rerum, en mode hongrois. 6. Marche funèbre.
7. Sursum corda.
     1-4, 1877. 5, 1872. 6, 1867. 7, 1877.

164. Albumblatt in E. *C.* 1841; from *210*.

165. Feuilles d'album in A flat. 1841.   II, 10

166. Albumblatt in waltz form. 1842. Cf. *212*.   II, 10

167. Feuille d'album in A minor. *C.* 1843; from *274*.

168. Elégie sur des motifs du Prince Louis Ferdinand de Prusse
1842.

169. Romance. 1848; cf. *132, 527*. From *301a:*

170. Ballade No. 1 in D flat. 1845-8.   II, 8

171. Ballade No. 2 in B minor. 1853.   II, 8

171a. Madrigal. (Earlier version of *172*, 5).

172. Consolations. 1849-50.   II, 8

173. Harmonies poétiques et religieuses.   II, 7
   1. Invocation. 2. Ave Maria. 3. Bénédiction de Dieu dans
la solitude. 4. Pensée des morts. 5. Pater noster. 6. Hymne
de l'enfant à son reveil. 7. Funérailles, Oct. 1849. 8.
Miserere, d'après Palestrina. 9. Andante lagrimoso. 10.
Cantique d'amour.
     3 sketched 1845; the rest 1847-52; 2 from *20*, 4 from *154*
and *691*; 5 from *21*, 2; 6 from *19*.

174. Berceuse; 1st version, 1854; 2nd version 1862.   II, 9

175. Légendes. 1. St. Francois d'Assise. La Prédication aux
                   oiseaux.   II, 9
     2. St. Francois de Paule marchant sur les flots.
   1863, at latest; cf. *4*, *28*, *354*.

176. Grosses Konzertsolo. *C.* 1849. Cf. *258, 365*.   II, 8

177. Scherzo and March. 1851.   II, 8

178. Sonata in B minor. 1852-3.   II, 8

179. Weinen, Klagen, Sorgen, Zagen; Prelude after J. S. Bach.
1859; cf. *180*.   II, 9

180. Variations on the theme of Bach—Basso continuo of the first
movement of his cantata Weinen, Klagen, Sorgen, Zagen,
and of the Crucifixus of the B minor Mass. 1862.   II, 9

181. Sarabande and Chaconne from Handel's opera Almira. 1879.

182. Ave Maria for the piano school of Lebert and Stark (The Bells
of Rome). 1862.   II, 9
183. Alleluja et Ave Maria (d'Arcadelt). 1862; 1 on themes from *4.*
184. Urbi et orbi. Bénédiction papale. 1864.
185. Vexilla regis prodeunt. 1864.
186. Weihnachtsbaum. Arbre de Noel. 1874-6.   II, 9
    1. Psallite. 2. O heilige Nacht! 3. Die Hirten an der Krippe.
    4. Adeste fideles. 5. Scherzoso. 6. Carillon. 7. Schlummer-
    lied. 8. Altes provençalisches Weihnachtslied. 9. Abend-
    glocken. 10. Ehemals! (Jadis). 11. Ungarisch. 12. Polnisch.
187. Sancta Dorothea. 1877.   II, 9
188. In festo transfigurationis Domini nostri Jesu Christi. 1880.
                                                                  II, 9
189. Piano piece in A flat. 1866.
190. La Marquise de Blocqueville. Portrait en musique. 1868.
191. Impromptu. 1872. (Also new version as Nocturne).   II, 9
192. Five little piano pieces (for Baroness von Meyendorff).   1, 2,
    1865; 3, 1873; 4, 1876; 5, (Sospiri) 1879. L.S. I; II, 10
    1, cf. *308.*   (1-4 only)
193. Klavierstuck in F sharp major. (In L.'s late years.)   II, 10
194. Mosonyis Grabgeleit. Mosonyi gyázmenete. 1870. L.S. III
195. Dem Andenken Petőfis. Petőfi Szellemenek. 1877; cf. *349.*
196. (First) Elegy. 1874, from *130.*   L.S. III; II, 9
197. Second Elegy. 1877; cf. *131.*   L.S. III; II, 9
197a. Toccata. 1879.
198. Wiegenlied (Chant du herceau). 1881; cf. *107, 133.*
199. Nuages gris. 1881.   L.S. I; II, 9
200. La lugubre gondola. (2 versions.) 1882; cf. *134.* L.S. I; II, 9
201. R.W.—Venezia. 1883.   L.S. I; II, 9
202. Am Grabe Richard Wagners. 1883; cf. *135, 267.*   L.S. II
203. Schlaflos, Frage und Antwort (Nocturne after a poem by
    Toni Raab). 1883.   L.S. III; II, 9
204. Recueillement. (In L.'s last years.)   II, 9
205. Historische ungarische Bildnisse. Magyar arcképek. 1885.
    1. Stephen Szechenyi.   2. Josef Eötvös.   3. Michael
    Vörösmarty.   4. Ladislaus Teleky.   5. Franz Deák.
    6. Alexander Petőfi.   7. Michael Mosonyi.
    1 is identical with *206;* 6 and 7 are *195* and *194.*
206. Trauervorspiel und Trauermarsch. 1885.   L.S. I;
                                                  II, 9 (Prelude only)
207. En Rêve. Nocturne.   L.S. I; II, 9
208. Unstern. Sinistre. Disastro. (In L.'s last years.) L.S. I; II, 9

WORKS IN DANCE FORMS

209. Grande Valse di Bravura (1st version). 1836; cf. *214*, 1. II, 10
210. Valse mélancolique (1st version). 1839. Cf. *214*, 2. II, 10
211. Ländler in A flat. 1843.
212. Petite Valse favorite. 1842. Cf. *166*, *213*.
213. Valse Impromptu. *C.* 1850, from *212*. II, 10
214. Trois Caprices—Valses. *C.* 1850; 1 from *209*, 2 from *210*,
    3 from *401*.   2, 3, L.S. IV; II, 10 (1 and 2 only)
214a. Carousel de Mme. Pelet-Narbonne. 1879.
215. 4 valses oubliées. 1881-5. 2, 3, L.S. IV; II, 10 (1-3 only)
216. Mephisto Waltz No. 3. 1883. II, 10
216a. Bagatelle sans Tonalité. 1885.
217. Mephisto Polka. 1883. II, 10
218. Galop in A minor. 1841(?) II, 10
219. Grand Galop chromatique. 1838. II, 10
220. Galop de Bal. *C.* 1840.
221. Mazurka brillante. 1850. II, 10
222. Mazurka in A flat.
223. 2 Polonaises. 1, C minor. 2, E major. 1851. II, 10
224. Csárdás macabre. 1881-2. L.S. I
225. 2 Csárdás. 1, Allegro. 2, Csárdás obstiné. 1884. 2, L.S. III
226. Festvorspiel—Prelude. 1856, as Preludio pomposo.
227. Festmarsch zur Saecularfeier von Goethe's Geburtstag (1st
    version). 1849; cf. *115*. II, 10
228. Huldigungsmarsch. 1853.
229. Vom Fels zum Meer. Deutscher Siegesmarsch. 1853-6.
230. Bülow-Marsch. 1883. II, 10
231. Heroischer Marsch im ungarischen Styl. 1840. Cf. *103*.
232. Seconde Marche hongroise. Ungarischer Sturmmarsch (1st
    version). 1844; cf. *119*.
233. Ungarischer Geschwindmarsch. Magyar Gyors indulo. 1870.

WORKS ON NATIONAL THEMES
Austrian
233a. Tyrolean Melody. 1856 (?) cf. *733*.
Czech
234. Hussitenlied of the 15th century. 1840.

English
235. God Save the Queen. 1841.
French
236. Faribolo Pastour and Chanson du Béarn. 1844.
237. La Marseillaise. Cf. *690*. Pub. 1872.
238. La cloche sonne. *C.* 1850.
239. Vive Henri IV. *C.* 1870-80.

German

240. Gaudeamus igitur. Concert paraphrase. 1843; separate work
from *71.*

Hungarian

241. Two movements of Hungarian character (Zum Andenken).
1828.

242. A. Magyar Dallok. Ungarische Nationalmelodien. 1839-47.
    Book   I. 1. Lento. 2. Andantino. 3. Sehr langsam.
                 4. Animato (cf. *243,* 2, and *244,* 6). 5.
                 Tempo giusto (cf. *243,* 1, and *244,* 6).
                 6. Lento (cf. *244,* 5).
    Book   II. 7. Andante cantabile (cf. *244,* 4).
    Book III. 8. Lento. 9. Lento.
    Book IV. 10. Adagio sostenuto a capriccio (cf. *244,*
                 15). 11. Andante sostenuto (cf. *243,* 3,
                 and *244,* 3 and 6).
  B. Magyar Rhapsodiák. Rhapsodies Hongroises.
    Book   V. 12. Mesto (heroide elégiaque) (cf. *244,* 5).
    Book VI. 13. Tempo di Marcia (cf. *244,* 15).
    Book VII. 14. Lento a capriccio (cf. ʾ*244,* 11).
    Book VIII. 15. Lento, Tempo e stilo zingarese   (cf.
                 244, 7).
    Book IX. 16. E major (cf. *244,* 10).
    Book   X. 17. Andante sostenuto (cf. *244,* 13).
  C.   18. Adagio (cf. *244,* 12). 19. Lento patetico
                 (cf. *244,* 8). 20. Allegro vivace (cf. *244,*
                 6 and 12). 21. Tempo di marcia funèbre
                 (cf. *244,* 14).

243. Ungarische Nationalmelodien. 1 from *242,* 5; 2 from *242,* 4;
3 from *242,* 11. *C.* 1840. Cf. *244,* 6.

244. Hungarian Rhapsodies.   II, 12
    1. Lento, quasi Recitativo. 1846.
    2. Lento a capriccio. 1847.
    3. Andante. From *242,* 11.
    4. Quasi Adagio, altieramente. From *242,* 7.
    5. Heroide-Elégiaque. Lento con duolo. From *242,* 12;
       cf. *242,* 6.
    6. Tempo giusto. From *242,* 4, 5, 11, 20; cf. *243.*
    7. Lento. From *242,* 15.
    8. Lento a capriccio. From *242,* 19.
    9. Pester Karneval. Moderato (2 versions). 1848.
    10. Preludio. From *242,* 16.

11. Lento a capriccio. From *242*, 14.
12. Mesto. From *242*, 18 and 20.
13. Andante sostenuto. From *242*, 17.
14. Lento quasi marcia funèbre. From *242*, 21. Cf. *123*.
15. Rákóczy March. 1st version from *242*, 10 and 13. 2nd version from *117*. Simplified version 1852, from 1st version.
16. Allegro. 1882.
17. Lento. *C.* 1883(?).
18. Lento. 1885.
19. Lento. 1885, after the Csárdás nobles of Ábrányi.

245. 5 Hungarian folk songs, transcribed for pf. *C.* 1873. L.S. III
246. Puszta-Wehmut. A Puszta Keserve. (In L.'s last years.)

Italian

247. La romanesca. 1839.
248. Canzone Napolitana. 1842.

Polish

249. Glanes de Woronince. 1. Ballade d'Ukraine (Dumka). 2. Mélodies polonaises. 3. Complainte (Dumka). 1847-8; 2, cf. *480*, 1.

Russian

250. Deux Mélodies russes. Arabesques. 1. Le Rossignol, air russe d'Alabieff. 2. Chanson bohémienne. 1842.
251. Abschied. Russisches Volkslied. 1885.

Spanish

252. Rondeau fantastique sur un thème espagnol, El Contrabandista. 1836.
253. Grosse Konzertfantasie über Spanische Weisen. 1845.
254. Rhapsodie espagnole. Folies d'Espagne et Jota aragonesa. *C.* 1863; one theme in common with *253*.

8. PIANOFORTE, FOUR HANDS

255. Festpolonaise. 1876, for the marriage of Princess Marie of Saxony.
256. Variation on the Chopsticks theme. 1880, for the collection of Borodin, Cui, Liadov and Rimsky-Korsakov.
256a. Notturno (F sharp major). From *158*.

9. Two Pianofortes

257. Grosses Konzertstuck über Mendelssohns Lieder ohne Worte. 1834.

258. Concerto pathétique. 1856 (at latest) from *176.*

10. Organ

259. Fantasy and Fugue on the chorale Ad nos, ad salutarem undam. 1850; theme from Meyerbeer's Le Prophète. Cf. *414.*

260. Prelude and Fugue on the name BACH. 1st version, 1855; 2nd version, 1870. Cf. *529.*

261. Pio IX. Der Papsthymnus. 1863(?). Later became *3, 8.*

261a. Andante religioso.

262. Ora pro nobis. Litanei. 1864.

263. Resignazione. 1877.

264. Missa pro organo lectarum celebrationi missarum adjumento inserviens. 1879, from *8* and *20.*

265. Gebet. 1879.

266. Requiem für die Orgel. 1883, from *12.*

267. Am Grabe Richard Wagners. 1883. Cf. *135, 202.*

268. Zwei Vortragsstücke. 1. Introitus. 2. Trauerode (Les Morts, *112,* 1). 1, 1884; 2, 1860.

11. Songs

269. Angiolin dal biondo crin (Bocella). (i) 1839; (ii) *c.* 1856.
VII, 2

270. Tre Sonetti di Petrarca.   VII, 2
    1. Pace non trovo (No. 104).
    2. Benedetto sia 'l giorno (No. 47).
    3. I vidi in terra angelici costumi (No. 123).
    1st version 1838-9; cf. *158.* 2nd version 1861; 1 and 2 in reverse order; cf. *161,* 4-6.

271. Il m'aimait tant (Delphine Gay). *C.* 1840.   VII, 1

272. Im Rhein (Heine). (i) *C.* 1840; (ii) *c.* 1856.   VII, 1, 2

273. Die Loreley (Heine). (i) 1841; (ii) *c.* 1856.   VII, 2

274. Die Zelle in Nonnenwerth (Lichnowsky). (i) 1841; (ii) 1860. Cf. *167.*   VII, 3

275. Mignons Lied (Goethe). (i) 1842; (ii) *c.* 1856; (iii) 1860. VII, 2

276. Comment, disaient-ils (Hugo). (i) 1842; (ii) *c.* 1859. VII, 1, 2

277. Bist du (Metschersky). 1843; rev. 1877-8.   VII, 3

278. Es war ein König in Thule (Goethe). (i) 1842; (ii) *c.* 1856.
VII, 2

279. Der du von dem Himmel bist (Goethe). (i) 1842; (ii) *c.* 1856; (iii) *c.* 1860; (iv) in L.'s last years.   VII, 1, 2

280. Freudvoll und leidvoll (Goethe).
    1st setting. (i) *C.* 1844; (ii) *c.* 1860.   VII, 1, 2
    2nd setting. *C.* 1848.   VII, 1
281. Die Vätergruft (Uhland). 1844.   VII, 2
282. O quand je dors (Hugo). (i) 1842; (ii) *c.* 1859.   VII, 1, 2
283. Enfant, si j'étais roi (Hugo). (i) *C.* 1844; (ii) *c.* 1859. VII, 1, 2
284. S'il est un charmant gazon (Hugo). (i) *C.* 1844; (ii) *c.* 1859.
      VII, 1, 2
285. La tombe et la rose (Hugo). *C.* 1844.   VII, 1
286. Gastibelza, Bolero (Hugo). *C.* 1844.   VII, 1
287. Du bist wie eine Blume (Heine). *C.* 1842-3.   VII, 2
288. Was Liebe sei (Charlotte von Hagn).
    1st setting *c.* 1843.   VII, 1
    2nd setting *c.* 1855.   VII, 2
    3rd setting *c.* 1878.   VII, 3
289. Vergiftet sind meine Lieder (Heine). 1842.   VII, 2
290. Morgens steh' ich auf und frage (Heine). (i) *C.* 1843; (ii)
    c. 1855.   VII, 1, 2
291. Die tote Nachtigall (Kaufmann). (i) *C.* 1843; (ii) 1878.
      VII, 1, 3
292. Songs from Schiller's Wilhelm Tell. 1. Der Fischerknabe.
    2. Der Hirt. 3. Der Alpenjäger. (i) *C.* 1845; (ii) *c.* 1859.
      VII, 1, 2
293. Jeanne d'Arc au bûcher (Dumas). (i) 1845; (ii) 1874.   VII, 3
294. Es rauschen die Winde (Rellstab). (i) *C.* 1845; (ii) *c.* 1860.
      VII, 2
295. Wo weilt er? (Rellstab). *C.* 1845.   VII, 2
296. Ich möchte hingehn (Herwegh). *C.* 1845.   VII, 2
297. Wer nie sein Brot mit Tränen ass (Goethe). 1st setting *c.* 1845;
    2nd setting *c.* 1860.   VII, 2, 3
298. O lieb, so lang du lieben kannst (Freiligrath). *C.* 1845;
    cf. *541.*   VII, 2
299. Isten veled (Horvath). 1846-7.   VII, 3
300. Le juif errant (Béranger). 1847.
301. Kling leise, mein Lied (Nordmann). (i) 1848; (ii) *c.* 1860.
      VII, 1, 2
301a. Oh pourquoi donc (Mme. Pavloff). 1848; cf. *169.*
302. Die Macht der Musik (Duchess Helen of Orleans). 1848-9.
      VII, 1
303. Weimars Toten, Dithyrambe (Schober). 1848.   VII, 1
304. Le vieux vagabond (Béranger). 1848; foreshadows *97* and
    109.   VII, 1

305. Schwebe, schwebe, blaues Auge (Dingelstedt).   (i) 1845;
(ii) *c.* 1860.   VII, 1, 2

306. Über allen Gipfeln ist Ruh (Goethe). (i) *C.* 1848; cf. *75;*
(ii) *c.* 1859.   VII, 2

307. Hohe Liebe (Uhland). *C.* 1849; cf. *541.*   VII, 2

308. Gestorben war ich (Uhland). *C.* 1849; cf. *192,* 1, *541.* VII, 2

309. Ein Fichtenbaum steht einsam (Heine). 1st setting *c.* 1855;
2nd setting *c.* 1860.   VII, 2

310. Nimm einen Strahl der Sonne (Ihr Auge) (Rellstab). *C.* 1855.
VII, 2

311. Anfangs wollt' ich fast verzagen (Heine). 1856; rev. *c.* 1880.
VII, 2

312. Wie singt der Lerche schön (Fallersleben). *C.* 1856.   VII, 2

313. Weimars Volkslied (Cornelius). 1857; cf. *87.*   VII, 2

314. Es muss ein Wunderbares sein (Redwitz). 1857.   VII, 2

315. Ich liebe dich (Rückert). 1857.   VII, 3

316. Muttergottes-Sträusslein zum Mai-Monate (Müller). 1. Das
Veilchen. 2. Die Schlüsselblumen. 1857.   VII, 2

317. Lasst mich ruhen (Fallersleben). *C.* 1858.   VII, 2

318. In Liebeslust (Fallersleben). *C.* 1858.   VII, 2

319. Ich scheide (Fallersleben). 1860.   VII, 2

320. Die drei Zigeuner (Lenau). 1860.   VII, 3

321. Die stille Wasserrose (Geibel). Prob. 1860.   VII, 3

322. Wieder möcht' ich dir begegnen (Cornelius). 1860.   VII, 3

323. Jugendglück (Pohl). *C.* 1860.   VII, 3

324. Blume und Duft (Hebbel). *C.* 1860.   VII, 3

325. Die Fischerstochter (Coronini). 1871.   VII, 3

326. La Perla (Princess Therese von Hohenlohe). 1872.   VII, 3

327. J'ai perdu ma force et ma vie (Tristesse) (de Musset). 1872.
VII, 3

328. Ihr Glocken von Marling (Kuh). 1874.   VII, 3

329. Und sprich (Biegeleben). 1874; rev. 1878.   VII, 3

330. Sei still (Nordheim). 1877.   VII, 3

331. Gebet (Bodenstedt). *C.* 1878.   VII, 3

332. Einst (Bodenstedt). *C.* 1878.   VII, 3

333. An Edlitam (Bodenstedt). *C.* 1878.   VII, 3

334. Der Glückliche (Wilbrandt). *C.* 1878.   VII, 3

335. Go not, happy day (Tennyson). 1879.   VII, 3

336. Verlassen (Michell). 1880.   VII, 3

337. Des Tages laute Stimmen schweigen (von Saar). 1880. VII, 3

338. Und wir dachten der Toten (Freiligrath). *C.* 1884.   VII, 3

339. Ungarns Gott. A magyarok Istene (Petöfi). 1881.   VII, 3

340. Ungarisches Königslied. Magyar Király-dal (Ábrányi). Cf. *93.* 1883. VII, 3

12. OTHER VOCAL WORKS

341. Ave Maria IV. Voice and organ (or harm. or pf.). 1881.
342. Le Crucifix (Hugo). Alto and pf. or harm. 1884, in 3 versions.
V, 6
343. Sancta Caecilia. Alto and organ or harm. (In L.'s last years.)
V, 6
344. O Meer im Abendstrahl (Meissner). SA and pf. or harm.
C. 1880. VII, 3
345. Wartburg-Lieder from Der Braut Willkomm auf Wartburg
(Scheffel). 1872. VII, 3
1. Introduction and mixed chorus. 2. Wolfram
von Eschenbach. 3. Heinrich von Ofterdingen. 4.
Walther von der Vogelweide. 5. Der tugendhafte
Schreiber. 6. Biterolf und der Schmied von Ruhla. 7.
Reimar der Alte. (Soloists, chorus and orch.)

13. RECITATIONS

346. Lenore (Bürger) with pf. 1858. VII, 3
347. Vor hundert Jahren (Halm) with orch. 1859.
348. Der traurige Mönch (Lenau) with pf. 1860. VII, 3
349. Des toten Dichters Liebe. A holt költö szerelme (Jókai) with
pf. 1874; cf. *195.* VII, 3
350. Der blinde Sänger (A. Tolstoy) with pf. 1875. VII, 3

B. ARRANGEMENTS, TRANSCRIPTIONS, FANTASIES, etc.

14. ORCHESTRAL WORKS

Bülow
351. Mazurka-Fantasie. Orch. 1865.
Cornelius
352. Second Overture to The Barber of Bagdad. Completed by L.
from C.'s sketches, 1877.
Egressy and Erkel
353. Szózat and Hymnus. Arr. 1870-3.
Liszt
354. Two Legends, *175.* Orch. 1863.
355. Vexilla regis prodeunt, *185.* 1864.

356. Festvorspiel, *226.* 1857.   I, 11
357. Huldigungsmarsch zur Huldigungsfeier S.K.H. des Grossher-
       zogs Carl Alexander von Sachsen-Weimar, *228.* 1857. I, 11
358. Vom Fels zum Meer. Deutscher Siegesmarsch, *229.*   I, 12
359. Hungarian Rhapsodies, *244,* arr. by the composer and Franz
       Doppler. 1. *244,* 14; 2. *244,* 12; 3. *244,* 6; 4. *244,* 2; 5. *244,* 5;
       6. *244,* 9.
360. À la chapelle Sixtine, *461.*
361. Der Papsthymnus, *261.*
362. Benedictus from the Hungarian Coronation Mass, *11,* for
       violin and orch. 1875.

Schubert
363. 4 Marches from Opp. 40, 54 and 121. 1859-60. Cf. *425-6.*

Zarembski
364. Danses galiciennes. 1881.

15. For Pianoforte and Orchestra

Liszt
365. Grand Solo de Concert, *176.* C. 1850.

Schubert
366. Fantasia in C major, Op. 15 (Wanderer), symphonic arrange-
       ment. C. 1851.

Weber
367. Polonaise brillante, Op. 72. C. 1851.

16. Songs with Orchestra

Korbay
368. 2 Songs. 1883.

Liszt
369. Die Loreley, *273,* 2. 1860.
370. Mignons Lied, *275,* 3. 1860.
371. Die Vätergruft, *281.* 1886.
372. Three Songs from Schiller's Wilhelm Tell, *292.* C. 1855; later
       revised.
373. Jeanne d'Arc au bûcher, *293,* 2. 1858; rev. 1874.
374. Die drei Zigeuner, *320.* 1860.

Schubert
375. 6 Songs. 1. Die junge Nonne. 2. Gretchen am Spinnrade.
       3. Lied der Mignon. 4. Erlkönig. 5. Der Doppelgänger.
       6. Abschied. 1860.

376. Die Allmacht. T or S solo, male chorus and orch. 1871.

Géza Zichy

377. Der Zaubersee.

17. CHAMBER MUSIC, ETC.

377a. La Notte, *112*, 2. Violin and pf. 1864-6. Cf. *722*.

378. Angelus, *163*, 1. 1. Harmonium. 1877. 2. String quartet. 1880.

379. Pester Karneval, *244*, 9. Pf., violin and cello.

380. O du mein holder Abendstern from Tannhäuser, cf. *444*. Cello and pf. 1852.

381. Benedictus and Offertorium from the Hungarian Coronation Mass, *11*. Violin and pf. 1869.

382. Die Zelle in Nonnenwerth, *274*. Pf. and violin or cello.

383. Die drei Zigeuner, *320*. Violin and pf. 1864.

18. FOR PIANOFORTE SOLO

PARAPHRASES, OPERATIC TRANSCRIPTIONS, ETC.

Alabieff, see *250*

384. Mazurka pour piano composée par un *amateur de St. Petersbourg*, paraphrasée par F. L. 1842.

Auber

385. Grande Fantaisie sur la Tyrolienne de l'opéra La Fiancée. 1829.

386. Tarantelle di bravura d'après la Tarantelle de La Muette de Portici (Masaniello). 1846.

387. 2 pf. pieces on themes from La Muette de Portici (one on the Berceuse).

Beethoven

388. Capriccio alla turca sur des motifs de Beethoven (Ruines d'Athènes). 1846; on same theme as *389*.

389. Fantasia on Beethoven's Ruins of Athens. From *122*.

Bellini

390. Reminiscences des Puritains. 1836.

391. I Puritani. Introduction et Polonaise. 1840; Polonaise from *390*.

392. Hexameron, Morceau de Concert. Grandes Variations de Bravoure sur le Marche des Puritains. 1837; with Thalberg, Pixis, Herz, Czerny and Chopin.

393. Fantaisie sur des motifs favoris de l'opéra La Sonnambula. 1839.

394. Réminiscences de Norma. 1841.

Berlioz

395. L'Idée fixe. Andante amoroso. *C.* 1833; on theme of Symphonie Fantastique. Rev. 1846(?) and 1865.

396. Benediction et Serment, deux motifs de Benvenuto Cellini. 1852.

Donizetti (Gaetano)

397. Réminiscences de Lucia de Lammermoor. 1835-6 (the sextet).

398. Marche et cavatine de Lucie de Lammermoor. 1835-6; with *397.*

399. Nuits d'Été à Pausilippe. 1. Barcajuolo. 2. L'Alito di Bice. 3. La Torre di Biasone. 1838.

400. Réminiscences de Lucrezia Borgia. 1. Trio du second acte. 2. Fantaise sur des motifs favoris de l'opéra; Chanson à boire (Orgie)—Duo—Finale. 1840.

401. Valse a capriccio sur deux motifs de Lucia et Parisina. 1842 (1st version; cf. *214*).

402. Marche funèbre de Dom Sébastien. 1844.

Donizetti, Giuseppe

403. Grande Paraphrase de la marche de Donizetti composée pour Sa Majesté le sultan Abdul Medjid-Khan. 1847.

E.H.z.S.-C.-G. (Ernst Herzog zu Sachsen-Coburg-Gotha)

404. Halloh! Jagdchor und Steyrer from the opera Tony. 1849.

Erkel, Franz

405. Schwanengesang and March from Hunyadi Laszlo. 1847.

Glinka

406. Tscherkessenmarsch from Russlan and Ludmila. 1843.

Gounod

407. Valse de l'opéra Faust. 1861.

408. Les Sabéennes. Berceuse de l'opéra La Reine de Saba. *C.* 1865.

409. Les Adieux. Rêverie sur un motif de l'opéra Romeo et Juliette. *C.* 1868.

Halévy

409a. Réminiscences de La Juive. 1835.

Mendelssohn

410. Wedding March and Dance of the Fairies from the music to Shakespeare's A Midsummer Night's Dream. 1849-50.

Mercadante

411. Soirées italiennes. Six amusements. 1838. 1. La primavera. 2. Il galop. 3. Il pastore svizzero. 4. La serenata del marinaro. 5. Il Brindisi. 6. La zingarella spagnola.

Meyerbeer

412. Grande Fantaisie sur des thèmes de l'opéra Les Huguenots. 1836.

413. Réminiscences de Robert le Diable. Valse infernale. 1841.

414. Illustrations du Prophète. 1849-50.
 1. Prière, Hymne triomphale, Marche du sacre. 2. Les Patineurs, Scherzo. 3. Choeur pastoral, Appel aux armes. (4 is *259.*)

415. Illustrations de l'Africaine. 1. Prière des matelots. 2. Marche indienne. 1865.

416. Le Moine. 1841; contains two other themes of Meyerbeer's.

Mosonyi, Michael

417. Fantaisie sur l'opéra hongrois Szép Ilonka. 1867.

Mozart

418. Réminiscences de Don Juan. 1841.

Pacini

419. Divertissement sur la cavatine I tuoi frequenti palpiti (Niobe). 1835-6.

Paganini

420. Grande Fantaisie de bravoure sur la Clochette. 1831-2. Cf. *140,* 3.

Raff

421. Andante Finale and March from the opera King Alfred. 1853.

Rossini

422. La Serenata e l'Orgia. Grande Fantaisie sur des motifs des Soirées musicales. 1835-6. Cf. *424,* 10 and 11 (also 1).

423. La Pastorella dell' Alpi e Li marinari. 2me Fantaisie sur des motifs des Soirées musicales. Cf. *424,* 6 and 12 (also 2). 1835-6.

424. Soirées musicales. 1837; cf. *422, 423.*
 1. La promessa. 2. La regata veneziana. 3. L'invito. 4. La gita in gondola. 5. Il rimprovero. 6. La pastorella dell' Alpi. 7. La partenza. 8. La pesca. 9. La danza. 10. La serenata. 11. L'orgia. 12. Li marinari.

Schubert

425. Mélodies hongroises (d'après Schubert). 1. Andante. 2. Marcia. 3. Allegretto. 1838-9, from the Divertissement à l'hongroise, Op. 54. Cf. *363.*

426. Schubert's Marches for the pianoforte. Nos. 1-3. 1846, from the 4-hand marches, Op. 40 and 121. Cf. *363.*

427. Soirées de Vienne. (9) Valses caprices after Schubert. 1852, from the waltzes, Op. 9, 18, 33, 50, 67, 77. Cf. *155*, 3. Cadenzas to 6 and 9, 1883.

Sorriano

428. Feuille morte. Elégie d'après Sorriano. *C.* 1845.

Tchaikovsky

429. Polonaise from Eugene Onegin. 1880.

Végh, Janos

430. Concert Waltz after the 4-hand waltz suite. (In L.'s last years.)

Verdi

431. Salve Maria de Jerusalem (I Lombardi). 1848.

431a. Concert Paraphrase on Ernani. 1847 (different from *432*).

432. Ernani. Paraphrase de concert. By 1849; rev. 1859.

433. Miserere du Trovatore. 1859.

434. Rigoletto. Paraphase du concert. 1859.

435. Don Carlos. Transcription. Coro di festa e marcia funebre. 1867-8.

436. Aïda. Danza sacra e duetto finale. 1871-9.

437. Agnus Dei de la messe de Requiem. 1877.

438. Réminiscences de Boccanegra. 1882.   L.S. II

Wagner

439. Phantasiestück on themes from Rienzi. 1859.   Arr. 1

440. Spinning chorus from The Flying Dutchman. 1860.   Arr. 1

441. Ballad from The Flying Dutchman. Before 1873.   Arr. 1

442. Overture to Tannhäuser. 1848.   Arr. 1

443. Pilgrims' chorus from Tannhäuser. 1st version *c.* 1861. 2nd version *c.* 1862(?).   Arr. 1

444. O du mein holder Abendstern.   Recitative and Romance from Tannhäuser, 1849. Cf. *380*.   Arr. 1

445. Two pieces from Lohengrin and Tannhäuser. 1852.   Arr. 1
    1. Entry of the guests on the Wartburg. 2. Elsa's bridal procession to the minster.

446. From Lohengrin. 1. Festival and bridal song. 2. Elsa's dream and Lohengrin's rebuke. 1854.   Arr. 1

447. Isolda's Liebestod from Tristan and Isolda. 1867.   Arr. 1

448. Am stillen Herd from Meistersinger. 1871.   Arr. 1

449. Valhalla from The Ring of the Nibelung. 1878-80.   Arr. 1

450. Solemn march to the Holy Grail from Parsifal. 1882.   Arr. 1

Weber

451. Freischütz Fantasy. 1840-1.

452. Leyer and Schwert. 1846-7.

453. Einsam bin ich, nicht alleine, from Preciosa. 1848.

454. Schlummerlied von C. M. von Weber mit Arabesken. 1848.

455. Polonaise brillante. C. 1851, from *367*.

Zichy, Count Géza

456. Valse d'Adele. Composée pour la main gauche seule. Transcription brillante à deux mains. C. 1877.

Unknown

458. Piano piece on Italian operatic melodies. (?)

459. Three short pieces on themes by other composers. (?)

460. Kavallerie-Geschwindmarsch. Pub. 1883.

PARTITIONS DE PIANO, TRANSCRIPTIONS, ETC.

Allegri and Mozart

461. À la chapelle Sixtine. Miserere d'Allegri et Ave verum corpus de Mozart. 1862; cf. *360*.

Arcadelt, see 183.

Bach

462. Six organ preludes and fugues. 1842-50.
    1. A minor. 2. C major (4-4). 3. C minor. 4. C major (9-8).
    5. E minor. 6. B minor.

463. Organ Fantasy and Fugue in G minor. Before 1872.

Beethoven

464. Symphonies de Beethoven. Partitions de Piano.   Arr. 2, 3
    5-7, 1837; Marcia funebre of 3, 1843; rest, and revision of
    the others, 1863-4.

465. Grand Septuor, Op. 20. 1841.

466. Adelaide, Op. 46. 1839; later rev.

467. 6 geistliche Lieder (Gellert), Op. 48. 1840.
    1. Gottes Macht und Versehung. 2. Bitten. 3. Busslied.
    4. Vom Tode. 5. Die Liebe des Nächsten. 6. Die Ehre
    Gottes aus der Natur.

468. Beethoven's Lieder von Goethe (from Op. 75, 83, 84). Before
    1849.
    1. Mignon. 2. Mit einem gemalten Bande. 3. Freudvoll und
    leidvoll. 4. Es war einmal ein König. 5. Wonne der
    Wehmut. 6. Die Trommel gerühret.

469. An die ferne Geliebte. Song-cycle, Op. 98. 1849.

Berlioz

470. Episode de la vie d'un artiste. Grande Symphonie Fantastique. Partition de Piano. 1833; new transcription of
    Marche au Supplice. 1864-5. Cf. *395*.

471. Ouverture des Francs-Juges. 1833.

472. Harold en Italie. Partition de piano (avec la partie d'alto).
1836.

473. Marche des Pèlerins de la sinfonie Harold en Italie. Transcrite
pour le piano. *C.* 1836.

474. Ouverture du Roi Lear. 1836.

475. Danse des Sylphes de la Damnation de Faust. *C.* 1860.

Bertin, Louise

476. Esmeralda. Opéra en 4 actes. Accompagnement de piano.
1837.

477. Air chanté par Massol (from Esmeralda) arrangé pour le
piano. 1837.

Bulhakov

478. Russischer Galopp. 1843.

von Bülow

479. Dante's Sonnet, Tanto gentile e tanto onesta. 1874.

Chopin

480. 6 Chants polonais, Op. 74. Between 1847 and 1860. Cf. *249,* 2.
1. Mädchens Wünsch. 2. Frühling. 3. Das Ringlein.
4. Bacchanal. 5. Meine Freuden. 6. Heimkehr.

Conradi

481. Le célèbre Zigeunerpolka. *C.* 1847.

Cui

482. Tarantella. 1885.

Dargomizhsky

483. Tarantelle, transcrite et amplifiée pour le piano à deux mains.
1879.

David, Ferdinand

484. Bunte Reihe, Op. 30. 1850.

Dessauer

485. Lieder. 1. Lockung. 2. Zwei Wege. 3. Spanisches Lied. 1847.

Draeseke

485a. Cantata, Der Schwur am Rütli, Pt. I. Pf. reduction. 1870.

Egressy and Erkel

486. Szózat und Ungarischer Hymnus. Not before 1870; cf. *353.*

Festetics, Count Lee

487. Spanisches Ständchen. 1846.

Franz

488. Er ist gekommen in Sturm und Regen. 1848.

489. Lieder. (Pub. 1849.)
  I. Schilflieder, Op. 2.
    1. Auf geheimen Waldespfaden. 2. Drüben geht die
    Sonne scheiden. 3. Trübe wird's. 4. Sonnenunter-
    gang. 5. Auf dem Teich.
  II. 3 Lieder, from Op. 3 and 8.
    6. Der Schalk. 7. Der Bote. 8. Meeresstille.
  III. 4 Lieder, from Op. 3 and 8.
    9. Treibt der Sommer. 10. Gewitternacht. 11. Das
    ist ein Brausen und Heulen. 12. Frühling und Liebe.

Goldschmidt, Adalbert von
490. Liebesszene und Fortunas Kugel from Die Sieben Todssünden.
  1880.

Gounod
491. Hymne à St. Cecile. 1866.

Herbeck
492. Tanzmomente. 1869.

Hummel
493. Septet, Op. 74. 1848.

Lassen
494. Löse Himmel meine Seele. 1861.
495. Ich weil' in tiefer Einsamkeit. (Pub. 1872.)
496. From the music to Hebbel's Nibelungen and Goethe's Faust.
  I. Nibelungen. 1. Hagen und Krimhild. 2. Bechlarn.
  II. Faust. 1. Osterhymne. 2. Hoffest. Marsch und
    Polonaise. 1878-9.
497. Symphonisches Zwischenspiel (Intermezzo) zu Calderons
  Schauspiel, Über allen Zauber Liebe. Not before 1882.

Lessmann
498. 3 songs from J. Wolff's Tannhäuser. C. 1882.
  1. Der Lenz ist gekommen. 2. Trinklied. 3. Du schaust
  mich an.

Liszt
499. Cantico del Sol di San Francesco, *4.* 1881.
500. Excelsior! Preludio zu den Glocken des Strassburger Münsters,
  6. C. 1875.
501. From the Hungarian Coronation Mass, *11.* 1. Benedictus.
  2. Offertorium. 1867.
502. Weihnachtslied II, *32.* C. 1864.
503. Slavimo Slavno Slaveni!, *33.* C. 1863.
504. Ave Maria II, *38.* 1st version *c.* 1870, 2nd version *c.* 1872.

505. Zum Haus der Herrn ziehen wir, *671* (prelude to *57*).

506. Ave maris stella, *34*, 2. *C.* 1868.

507. 70 bars on themes from the first Beethoven Cantata, *67*. *C.* 1847.

508. Pastorale. Schnitterchor aus dem Entfesselten Prometheus, *69*. 1861.

509. Gaudeamus igitur. Humoreske, *71* (not *240*). *C.* 1870.

510. Marche héroique, *82*. *C.* 1848.

511. Geharnischte Lieder, *90*, 4-6. Before 1861.

512. Von der Wiege bis zum Grabe, *107*. 1881.

513. Gretchen. 2nd movement of the Faust Symphony, *108*. 1874.

514. Der Tanz in der Dorfschenke (1st Mephisto Waltz), *110*, 2. 1859-60.

515. Second Mephisto Waltz, *111*. 1881.

516. Les Morts, *112*, 1. 1860. Cf. *699*.

517. Le Triomphe funèbre du Tasse, *112*, 3. 1866.

518. Salve Polonia, *113*. After 1863.

519. Deux Polonaises de l'oratorio St. Stanislaus, *688*. In the 1870s.

520. Künstlerfestzug zur Schillerfeier, *114*. 1857-60.

521. Festmarsch zur Goethejubiläumsfeier, *115* (2nd version). 1857.

522. Festmarsch nach Motiven von E.H. zu S.-C.-G., *116*. *C.* 1859.

523. Ungarischer Marsch zur Krönungsfeier in Ofen-Pest, *118*. 1870.

524. Ungarischer Sturmmarsch, *119* (2nd version). 1875. Cf. *232*.

525. Totentanz, *126* (ii). *C.* 1860-5.

526. Epithalam zu Eduard Reményis Vermählungsfeier, *129*. *C.* 1872.

527. Romance oubliée, *132*. 1880. Cf. *169*, *301a*.

528. Festpolonaise, *255*. 1876.

529. Fantasie und Fuge über das Thema BACH, *260*, 2nd version. 1871.

530. L'Hymne du Pape. Inno del Papa. Der Papsthymnus, *261*. *C.* 1864.

531. Buch der Lieder für Piano allein. *C.* 1843.
     1. Loreley, *273*, 1. 2. Am Rhein, *272*, 1. 3. Mignon, *275*, 1. 4. Es war ein König in Thule, *278*, 1. 5. Der du von dem Himmel bist, *279*, 1. 6. Angiolin dal biondo crin, *269*, 1.

532. Loreley, 2nd version, *273*. 1861.

533. Il m'aimait tant, *271*. *C.* 1843.

534. Die Zelle in Nonnenwerth. Elégie, *274*. *C.* 1843.

535. Comment, disaient-ils, *276*, 1. 1847(?).

536. O quand je dors, *282*, 1. 1847(?).

537. Enfant, si j'étais roi, *283*, 1. 1847(?).

538. S'il est un charmant gazon, *284*, 1. 1847(?).
539. La tombe et la rose, *285*. 1847(?).
540. Gastibelza, *286*. 1847.
541. Liebesträume, 3 Notturnos, *307, 308* and *298*. *C.* 1850.
542. Weimars Volkslied, *87*. 1857.
543. Ungarns Gott, *339*. 1881.
544. Ungarisches Königslied, *340*. 1883.
545. Ave Maria IV, *341*. 1881.
546. Der blinde Sänger, *350*. 1878.

Mendelssohn
547. Lieder, from Op. 19, 34 and 47. 1840.
  1. Auf Flügeln des Gesanges. 2. Sonntagslied. 3. Reiselied.
  4. Neue Liebe. 5. Frühlingslied. 6. Winterlied. 7. Suleika.
548. Wasserfahrt and Der Jäger Abschied from Op. 50. 1848.

Meyerbeer
549. Festmarsch zu Schillers 100-Jähriger Geburtsfeier. 1860.

Mozart
550. Confutatis and Lacrymosa from the Requiem. (Pub. 1865.)

Pezzini
551. Una stella amica. Mazurka. (?)

Rossini
552. Ouverture de l'opéra Guillaume Tell. 1838.
553. Deux Transcriptions. 1. Air du Stabat Mater (Cujus animam).
  2. La Charité. 1847.

Rubinstein
554. 2 Songs. 1. O! wenn es doch immer so bliebe. 2. Der Asra.
  C. 1880(?).

Saint-Saëns
555. Danse macabre, Op. 40. 1876.

Schubert
556. Die Rose (Heidenröslein). 1835; rev. 1838.
557. Lob der Tränen. 1838.
558. 12 Lieder. 1838.
  1. Sei mir gegrüsst. 2. Auf dem Wasser zu singen. 3. Du
  bist die Ruh. 4. Erlkönig. 5. Meeresstille. 6. Die junge
  Nonne. 7. Frühlingsglaube. 8. Gretchen am Spinnrade.
  9. Ständchen (Hark, hark!). 10. Rastlose Liebe. 11. Der
  Wanderer. 12. Ave Maria.
559. Der Gondelfahrer, Op. 28. 1838.

560. Schwanengesang. 1838-9.
    1. Die Stadt. 2. Das Fischermädchen. 3. Aufenthalt. 4. Am Meer. 5. Abschied. 6. In der Ferne. 7. Ständchen (Leise flehen). 8. Ihr Bild. 9. Frühlingssehnsucht. 10. Liebesbotschaft. 11. Der Atlas. 12. Der Doppelgänger. 13. Die Taubenpost. 14. Kriegers Ahnung.

561. Winterreise. 1839.
    1. Gute Nacht. 2. Die Nebensonnen. 3. Mut. 4. Die Post. 5. Erstarrung. 6. Wasserflut. 7. Der Lindenbaum. 8. Der Leyermann. 9. Täuschung. 10. Das Wirtshaus. 11. Der stürmische Morgen. 12. Im Dorfe.

562. Geistliche Lieder. 1840.
    1. Litaney. 2. Himmelsfunken. 3. Die Gestirne. 4. Hymne.

563. Sechs Melodien. 1846.
    1. Lebewohl. 2. Mädchens Klage. 3. Das Sterbeglöcklein. 4. Trockene Blumen. 5. Ungeduld (1st version). 6. Die Forelle (1st version). Cf. *564, 565.*

564. Die Forelle (2nd version). 1846.

565. Müllerlieder. 1846.
    1. Das Wandern. 2. Der Müller und der Bach. 3. Der Jäger. 4. Die böse Farbe. 5. Wohin? 6. Ungeduld (2nd version).

Schumann

566. Liebeslied (Widmung). 1848.

567. An den Sonnenschein and Rotes Röslein. (Pub. 1861.)

568. Frühlingsnacht (Überm Garten durch die Lüfte). (Pub. 1872.)

569. Lieder von Robert and Clara Schumann. (Pub. 1872).
    A. Robert, from Op. 79 and 98a.
        1. Weihnachtslied. 2. Die wandelnde Glocke. 3. Des Sennen Abschied. 5. Er ist's. 6. Nur wer die Sehnsucht kennt. 7. An die Türen will ich schleichen.
    B. Clara.
        1. Warum willst du andere fragen? 2. Ich habe in deinem Auge. 3. Geheimes Flüstern.

570. Provençalisches Minnelied. 1881.

Spohr

571. Die Rose. Romanze. (Pub. 1877.)

Szabady

572. Revive Szegedin. Marche hongroise transcrite d'après l'orchestration de J. Massenet. 1879.

Széchényi, Count Imre

573. Einleitung und Ungarischer Marsch. Bevezetés és magyar indulo. (Pub. 1878.)

Weber

574. Overture Oberon. 1843.
575. Overture Der Freischütz. 1846.
576. Jubelouverture. 1846.

Wielhorsky, Count Michael

577. Autrefois. Romance. 1843.

19. PIANOFORTE DUET

Field

577a. Nocturnes Nos. 1-9, 14, 18 and Nocturne Pastorale in E.

Liszt

578. 4 Pieces from St. Elisabeth, *2. C.* 1866.
    1. Prelude. 2. March of the Crusaders. 3. Storm. 4. Interlude.
579. 2 Orchestral pieces from Christus, *3.* 1866-73.
    1. Hirtenspiel. 2. Die heiligen drei Könige.
580. Excelsior! Cf. *6. C.* 1875.
581. Benedictus and Offertorium from the Hungarian Coronation Mass, *11.* 1869.
582. O Lamm Gottes, unschuldig, *50,* 5. 1878-9.
583. Via Crucis, *53.* 1878.
584. Festkantate zur Enthüllung des Beethoven-Denkmals in Bonn, *67.* 1845.
585. Pastorale. Schnitterchor aus dem Entfesselten Prometheus, *69.* 1861.
586. Gaudeamus igitur. Humoreske, *71.* 1870-2.
587. Marche héroique, *82. C.* 1848.
588. Weimars Volkslied, *87.* 1857.
589. Ce qu'on entend sur la montagne, *95.* 1874.
590. Tasso, *96. C.* 1858.
591. Les Préludes, *97. C.* 1858.
592. Orpheus, *98. C.* 1858.
593. Prometheus, *99. C.* 1858.
594. Mazeppa, *100.* 1874.
595. Festklänge, *101.* 1854-61.
596. Hungaria, *103.* 1874(?).
597. Hamlet, *104.* 1874.
598. From the Cradle to the Grave, *107.* 1881.
599. Two episodes from Lenau's Faust, *110.* 1861-2.
600. Second Mephisto Waltz, *111.* 1881.
601. Les Morts, *112,* 1. 1860.
602. La Notte, *112,* 2. 1866.

603. Le Triomphe Funèbre du Tasse, *112*, 3. 1866 and 1869.
604. Salve Polonia, *113*. 1863.
605. Künstlerfestzug zur Schillerfeier, *114.* 1859.
606. Festmarsch zur Goethejubiläumsfeier, *115.* C. 1858.
607. Festmarsch nach Motiven von E.H. zu S.-C.-G., *116. C.* 1859.
608. Rákóczy March, *117.* 1870.
609. Ungarischer Marsch zur Kronungsfeier in Ofen-Pest, *118.* 1870.
610. Ungarischer Sturmmarsch, *119.* 1875.
611. Epithalam, *129.* 1872(?).
612. Elegie, *130.* 1874.
613. Weihnachtsbaum, *186. C.* 1876-82.
614. Dem Andenken Petöfis, *195.* 1877.
615. Grande Valse di Bravura, *209.* 1836.
616. Grand Galop Chromatique, *219.* 1838.
617. Csárdás macabre, *224.* 1882.
618. Csárdás obstiné, *225*, 2. 1884(?).
619. Bülow-Marsch, *230. C.* 1883.
620. Hussitenlied, *234.* 1840.
621. Hungarian Rhapsodies (from the orchestral version, *359*). 1874.
      I. *244*, 14; 2. *244*, 12; 3. *244*, 6; 4. *244*, 2; 5. *244*, 5; 6. *244*, 9.
622. Rhapsody No. 16, *244*, 16. 1882.
623. Rhapsody No. 18, *244*, 18. 1885.
624. Fantasy and Fugue on Ad Nos, *259*.
625. Der Papsthymnus, *261. C.* 1865.
626. Ungarisches Königslied, *340. C.* 1884.
627. Sonnambula Fantasy, *393. C.* 1852.
628. Benediction et Serment de Benvenuto Cellini, *396.* 1853(?).
629. Tscherkessenmarsch, *406.* 1843.
630. Réminiscences de Robert le Diable, *413.* 1841-3.
631. Andante Finale and March from King Alfred, *421.* 1853 (?).
632. Vier Märsche von Franz Schubert, *425*, 2; *426.* After 1860, from *363.*
633. À la Chapelle Sixtine, *461. C.* 1865.
634. Grand Septuor de Beethoven, *465.*
      Mozart
634a. Adagio from The Magic Flute.

      Liszt   20. Two Pianofortes

635. Ce qu'on entend sur la montagne, *95. C.* 1854-7.
636. Tasso, *96. C.* 1854-6.
637. Les Préludes, *97. C.* 1854-6.
638. Orpheus, *98. C.* 1854-6.
639. Prometheus, *99.* 1855-6.

640. Mazeppa, *100.* 1855.

641. Festklänge, *101.* C. 1853-6.

642. Héroide Funèbre, *102.* C. 1854-6.

643. Hungaria, *103.* C. 1854-61.

644. Hamlet, *104.* C. 1858-61.

645. Hunnenschlacht, *105.* 1857.

646. Die Ideale, *106.* 1857-8.

647. A Faust Symphony, *108.* 1856; rev. 1860.

648. A Symphony to Dante's Divina Commedia, *109.* C. 1856-9.

649. Fantasy on themes from Beethoven's Ruins of Athens, *122.* After 1852.

650. Piano Concerto No. 1, *124.* 1853.

651. Piano Concerto No. 2, *125.* C. 1859.

652. Totentanz, *126* (ii). Between 1859-65.

653. Wanderer Fantasia, *366.* After 1851.

654. Hexameron, *392.* After 1837.

655. Réminiscences de Norma, *394.* After 1841.

656. Réminiscences de Don Juan, *418.* After 1841.

657. Beethoven's 9th Symphony, cf. *464.* 1851, at latest.

21. ORGAN

Allegri and Mozart

658. Evocation à la Chapelle Sixtine (*461*). C. 1862.

Arcadelt

659. Ave Maria, cf. *183*, 2. 1862.

Bach

660. Einleitung und Fuge aus der Motette 'Ich hatte viel Bekümmernis' und Andante 'Aus tiefer Not.' 1860.

661. Adagio from the 4th violin sonata. (?)

Chopin

662. Preludes, Op. 28, Nos. 4 and 9.

Lassus

663. Regina coeli laetare. 1865.

Liszt

664. Tu es Petrus, *3*, 8. 1867; cf. *261.*

665. San Francesco. Preludio per il Cantico del Sol di San Francesco, *4.* 1880.

666. Excelsior! Preludio zu den Glocken des Strassburger Münsters, *6.* After 1874.

667. Offertorium from the Hungarian Coronation Mass, *11.* After 1867.

668. Slavimo Slavno Slaveni!, *33.* 1863.

669. Zwei Kirchenhymnen. 1. Salve Regina. 2. Ave maris stella, *34.* After 1868.
670. Rosario, *56,* 1-3. 1879.
671. Zum Haus des Herren ziehen wir, Prelude to *57.* 1884.
672. Weimars Volkslied, *87.* 1865.
673. Weinen, Klagen Variations, *180.* 1863.
674. Ungarns Gott, *339. C.* 1881.

Nicolai
675. Kirchliche Festouverture über den Choral 'Ein feste Burg ist unser Gott.' 1852.

Wagner
676. Chor der jüngeren Pilger aus Tannhäuser, cf. *443.* 1st version 1860; 2nd version 1862.

22. ORGAN WITH OTHER INSTRUMENTS

677. Hosannah for organ and trombone. 1862, from *4.*
678. Offertorium and Benedictus from the Hungarian Coronation Mass, *11,* for violin and organ. Prob. 1871.
679. Aria 'Cujus animam,' from Rossini's Stabat Mater, for organ and trombone, cf. *553, 682.* In the 1860s.

23. VOCAL ARRANGEMENTS

680. Ave maris stella, *34.* Voice and pf. or harm. 1868.
681. Ave Maria II, *38.* Voice and organ or harm. 1869.   V, 6
682. Cujus animam from Rossini's Stabat Mater. T and organ, cf. *553, 679.*   V, 7
683. Serbisches Lied (Ein Mädchen sitzt am Meeresstrand), in Musik gesetzt von F. W. C. Fürst von Hohenzollern-Hechingen, mit Pianoforte-Begleitung von F. L.
684. Barcarole vénitienne de Pantaleoni, avec accompagnement de pianoforte par F. L. (Pub. 1842.)
685. Es hat geflammt die ganze Nacht (Grand Duchess Pavlovna), with piano accomp. by L.

24. RECITATION

Draeseke
686. Helges Treue (Strachwitz). 1860; arr. from the song.

25. UNFINISHED WORKS

687. Sardanapale. Opera in three acts. Text after Byron by Rotondi (?). 1846-51.

688. Die Legende vom heiligen Stanislaus. Oratorio. Text after
Lucien Siemienski by Princess Sayn-Wittgenstein (?),
Cornelius and K. E. Edler. 1873-85; Salve Polonia, *113,*
1863; De Profundis, *16,* 1881. Cf. *519.*

689. Singe, wem Gesang gegeben. Male chorus. *C.* 1847.

690. Revolutionary Symphony. Orchestra. 1830; rev. 1848.
Cf. *102.*

691. De Profundis. Psaume instrumental. Piano and orch. *C.* 1834.
Cf. *126* (i), *173,* 4.

692. Violin Concerto. 1860, for Reményi.

693. Two Hungarian piano pieces. 1. B flat minor. 2. D minor.
C. 1840.   L.S. III

694. Fantasia on English themes. *C.* 1840.

695. Piano piece in F major. 1843(?).

696. Mephisto Waltz No. 4. 1885.   L.S. II

697. Fantasia on themes from Figaro. *C.* 1843. (Completed by
Busoni 1912.)

698. La Mandragore. Ballade de l'opéra Jean de Nivelle de L.
Delibes. After 1880.

699. La Notte, *112,* 2. Piano. 1864-6.

700. Carnaval de Venise (Paganini) for piano.

701. Den Felsengipfel stieg ich einst hinan. Voice and pf.

APPENDIX

1. DOUBTFUL OR LOST

*(For references see appendix to catalogue in " Grove's Dictionary " and
Friedrich Schnapp, " Verschollene Kompositionen Franz Liszts " (see
Bibliography).)*

1. Sacred Choral Works

702. Tantum ergo. 1822. Lost.

703. Psalm 2 for T solo, mixed chorus and orchestra, 1851. (Prob.
part of *690.*)

704. Requiem on the death of the Emperor Maximilian of
Mexico. (?)

705. The Creation. (?)

706. Benedictus, for mixed chorus and organ. Pub. 1939, New
York. Doubtful.

707. Excelsior, *6,* 1, for M-S or Bar. solo, male chorus and pf.
? not arr. by L.

2. Secular Choral Works

708. Rinaldo (Goethe) for T solo, male chorus and pf. ? not by L.

708a. A patakhoz (Garai). 1846. Lost. (?)

3. For Orchestra

709. Boze cos Polske. Also for pf., *113.*

710. Funeral March. (?)

711. Csárdás macabre, *224.* (?)

712. Romance oubliée, *132,* for viola and orch. Not arr. by L.

4. For Piano and Orchestra

713. Two concertos. 1825. Lost.

714. Concerto in the Hungarian style. (?)

715. Concerto in the Italian style. (?)

716. Grande Fantaisie Symphonique in A minor. (?)

5. Other Instrumental Works

717. Trio. 1825. Lost.

718. Quintet. 1825. Lost.

719. The Seasons, for string quartet. (?)

720. Allegro moderato in E major for violin and piano. (?)

721. Violin prelude. (?)

722. La Notte, *112,* 2, for violin and piano. Now *377a.*

723. Tristia, from Vallée d'Obermann, *160,* 6, for piano, violin and cello. (?)

6. For Piano Solo

724. Rondo; Fantasia. 1824. Lost.

725. 3 Sonatas. 1825. Lost.

726. Étude in C major. (?)

726a. Technical Studies. Vol. 3. (?)

727. Prelude omnitonique. (?)

728. Sospiri (Companion piece to *192*). See *192.*

729. Ecce panis angelorum. (?)

730. In Memoriam. (?)

731. Valse elégiaque. (?)

732. 4me Valse oubliée. See *215.*

733. Ländler in G major. See *233a.*

734. Ländler in D major. (?)

735. Air cosaque. (?)

736. Kerepesi csárdás. ? not by L.

737. Trois morceaux en style de danse ancien hongrois. Pub. 1850; not by L.

738. Spanish Folk-Tunes. (?)

739. Overture, Coriolan (Beethoven). Prob. lost.

740. Overture, Egmont (Beethoven). (?)

741. Overture, Le carnaval Romain (Berlioz). (?)

742. Duettino (Donizetti). (?)

743. Soldiers' chorus from Gounod's Faust. Prob. lost.

744. Paraphrase on the fourth act of Kullak's Dom Sebastian. (?)

745. Funeral March. (?)

746. Andante maestoso. (?)

747. Poco Adagio from the Gran Mass. (?)

748. Overture, The Magic Flute (Mozart). (?)

749. Radovsky. Preussischer Armeemarsch No. 120. Pf. score.

750. Introduction et Variations sur une Marche du Siège de Corinth (Rossini), 1830. (?)

751. Nonetto e Mose (Rossini), Fantaisie. (?)

752. Gelb rollt (Rubinstein). (?)

753. Alfonso und Estrella (Schubert), Act I. Pf. score. 1850-1. Prob. lost.

754. 2a Mazurka di P. A. Tirindelli, variata da F.L. Prob. not by L.

7. For Piano, Four Hands

755. Sonata. 1825. Lost.

756. Mosonyis Grabgeleit, *194.* (?)

8. For Two Pianos

757. Le triomphe funèbre du Tasse, *112*, 3. (?)

9. For Organ

758. Symphonic poem, The Organ, after Herder. (?)

759. Consolation No. 4 in D flat, *172.* ? not arr. by L.

760. Cantico del Sol, *4.* Cf. *665.*

761. Chopin's Funeral March for organ, cello and piano. (?)

10. Songs with Piano

762. Air de Chateaubriand. (?)

763. Strophes de Herlossohn. (?)

764. Kränze pour chant. (?)

765. Glöckchen (Müller). ? companion piece to *316.*

765a. L'aube naît (Hugo). 1842. Lost.

766. Der Papsthymnus, *3*, 8, for S or T and pf. ? not arr. by L.

11. Other Vocal Works

767. Excelsior, *6*, 1, for voice and organ. ? not arr. by L.

12. Recitations with pf.

768. Der ewige Jude (Schubart). (?)

2. PLANNED

Operas

1842   Le Corsaire.

1845   Stage version of La Divina Commedia, text by Autran.

1846   Two Italian operas.
1846-54 Opera on Faust.
1847   Richard en Palestine (Scott).
1848   Spartacus.
1856-8 Hungarian opera, János or Jánko.
1858-9 Jeanne d'Arc.

Chorus

1845   Les Laboureurs, Les Matelots, Les Soldats (Lamennais),
        companion pieces to *81*.
1849   Oratorio after Byron's Heaven and Earth; text by Wagner.
1856   Two Masses.
1860-1 Liturgie catholique, Liturgie romaine.
1862   Manfred.
1869   St. Etienne, roi d'Hongrie, or Fire and Water.
1874   Longfellow's Golden Legend as a recitation (cf. *6*).
        Theodor Körner's Fiedler der hl. Cäcilia.
        Sketches for Psalms 14 and 15, a plainsong Magnificat,
        Les Djinns, and a Miserere.

Orchestra

1853   Music to The Tempest.
1861   The History of the World in Sound and Picture (after a
        series of pictures by Kaulbach): The Tower of Babel,
        Nimrod, Jerusalem, The Glory of Greece.
1865   Continuation to Hungaria, *103*.
1871   Two works of Chopin for pf. and orch., incl. Fantasy Op. 49.
1885   Ungarische Bildnisse, *205*, for orch.

Pianoforte

1850   Schubert's C major Symphony.
1860   Polonaise martiale.
1863   Transcription of Beethoven's Quartets.
1877   New ed. of Mendelssohn's Songs Without Words.
1880   Bach's Chaconne.
1883   Piano score of Rousseau's Devin du Village.
1886   Fantasia on Mackenzie's The Troubadour.
        Sketch for Somogyi Csárdás.

3. WORKS BY OTHER COMPOSERS, EDITED BY LISZT

Bach, Kompositionen für Orgel.   Revidiert und mit Beiträgen
    versehen von F.L. (includes *660* and *661*).
    Chromatic Fantasy.
    Three Preludes and Fugues in C sharp minor.
    (In "Anthologie Classique," Schlesinger, Berlin.)

Beethoven. Three piano concertos (Op. 37, 58, 73) edited and arranged for 2 pianos, with cadenzas, 1879.

Works for piano, 2 and 4 hands, ed. 1857.

Duos for piano with violin or cello, horn, flute or viola.

Trios for piano with violin (or clarinet) and cello.

Mass in C. Mass in D.

Piano quartets, ed. 1861.

Trios for string and wind instruments.

Menuet, revu par F. Liszt.

Chopin. Études, pub. 1877.

Clementi. Preludes et Exercices, corrigés et marqués au metronome par le jeune Liszt, suivis de douze de ses Études. Cf. *136.* Pub. 1825.

Field. 18 Nocturnes, redigés et accompagnés d'une preface, 1859.

Gottschalg. Repertorium für Orgel, Harmonium oder Pedal-Flügel. Bearbeitet unter Revision und mit Beiträgen von F. L. Cf. *260, 660-4, 667, 678.*

Handel. Fugue in E minor.

Hummel. Septet, Op. 74, for piano, flute, oboe, horn, viola, cello and bass. Also as quintet for piano, violin, viola, cello and bass.

Mädchenlieder. Unter Mitwirkung von Hoffmann von Fallersleben und Franz Liszt herausgegeben von A. Bräunlich und W. Gottschalg. Pub. 1851.

Schubert. Selected sonatas and solo piano pieces. (2 and 4 hands.) Ed. 1868-80.

Scarlatti. Cat's Fugue.

Viole. Gartenlaube. 100 études for pf.

Weber. Selected sonatas and solo piano pieces. Ed. 1868, 1870.

Paraphrase on the Invitation to the Dance, 1843.

6 pages of variants to the Konzertstuck.

4. LITERARY WORKS

Gesammelte Schriften. Ed. Lina Ramann. Pub. 1880-3.

Vol. I. F. Chopin (1852).

Vol. II, 1. Essays from the Revue et Gazette Musicale:

On the position of artists (1835). On future Church music (1834).

On popular editions of important works (1836).

On Meyerbeer's Les Huguenots (1837).

Thalberg's Grande Fantaisie, Op. 22, and Caprices, Op. 15 and 19 (1837).

To M. Fétis (1837).

R. Schumann's piano compositions, Op. 5, 11, 14 (1837).

Paganini; a Necrology (1840).

(This volume does not include the article on Alkan's Trois Morceaux dans le genre pathétique, Op. 15 (1837).)

Vol. II, 2. Essays of a Bachelor of Music (1835-40).

1-3. To George Sand. 4. To Adolphe Pictet. 5. To Louis de Ronchand. 6. By Lake Como (to Louis de Ronchand). 7. La Scala (to M. Schlesinger). 8. To Heinrich Heine. 9. To Lambert Massart. 10. On the position of music in Italy (to M. Schlesinger). 11. St. Cecilia (to M. d'Ortigue). 12. To Hector Berlioz.

Vol. III, 1.

Gluck's Orpheus (1854). Beethoven's Fidelio (1854). Weber's Euryanthe (1854). On Beethoven's music to Egmont (1854). On Mendelssohn's music to A Midsummer Night's Dream (1854). Scribe and Meyerbeer's Robert le Diable (1854). Schubert's Alfonso und Estrella (1854). Auber's Muette de Portici (1854). Bellini's Montecchi e Capuletti (1854). Boieldieu's Dame Blanche (1854). Donizetti's La Favorita (1854). Pauline Viardot-Garcia (1859). No Entr'acte Music! (1855). Mozart; on the occasion of his centenary festival in Vienna (1856).

Vol. III, 2. Richard Wagner.

Tannhäuser and the Song Contest on the Wartburg (1849). Lohengrin and its first performance at Weimar (1850). The Flying Dutchman (1854). The Rhinegold (1855).

Vol. IV.

Berlioz and his Harold Symphony (1855). Robert Schumann (1855). Clara Schumann (1855). Robert Franz (1855). Sobolewski's Vinvela (1855). John Field and his Nocturnes (1859).

Vol. V.

On the Goethe Foundation (1850). Weimar's September Festival in honour of the centenary of Karl August's birth (1857). Dornröschen; Genast's poem and Raff's music (1855). Marx and his book, The Music of the Nineteenth Century (1855). Criticism of criticism; Ulibishev and Serov (1858). A letter on conducting; a defence (1853).

Vol. VI. The Gypsies and their Music in Hungary (1859).

Vol. VII (not published) was to contain two more letters of 1837-8 and 1841, the Illustrations to Benvenuto Cellini (1838), and L's forewords to his musical works.

Published separately:
    De la fondation Goethe à Weimar, 1851.
    Lohengrin et Tannhäuser de R. Wagner, 1851.
    F. Chopin, 1852.
    Des Bohemiéns et de leur musique en Hongrie, 1859.
    Über John Field's Nocturne, 1859.
    R. Schumann's Musikalische Haus- und Lebensregeln. French
        translation by F. L. 1860.
    Robert Franz, 1872.

Liszt also took part, together with Marschner, Reissiger and Spohr, in Eduard Bernsdorf's Neues Universal-Lexicon der Tonkunst (1856-65). A manual of piano technique, written for the Geneva Conservatoire (1835 (?)) is apparently lost.

BIBLIOGRAPHY

BOOKS AND ARTICLES on Liszt are legion; I can only quote here those which are most likely to be of use to an English reader, together with the most important original sources in other languages. Those who wish for further bibliographical details are recommended to consult *Liszt Ferenc Bibliográfiai Kísérlet; Franz Liszt, ein Bibliographischer Versuch*, by Lájos Koch (Budapest 1936); this is exhaustive for all material published up to that time. It is written in Hungarian and German.

ORIGINAL SOURCES

Franz Liszts Briefe, ed. La Mara. 8 vols. Leipzig, 1893-4.

Franz Liszts Briefe an Baron Anton Augusz, ed. Wilhelm von Csapo. Budapest, 1911.

Franz Liszts Briefe an seine Mutter, ed. La Mara. Leipzig, 1918.

Franz Liszts Briefe an Carl Gille, ed. Adolf Stern. Leipzig, 1903.

Briefe hervorragender Zeitgenossen an Franz Liszt, ed. La Mara. Leipzig, 1895, 1904.

Briefwechsel zwischen Franz Liszt und Hans von Bülow, ed. La Mara. Leipzig, 1898.

Briefwechsel zwischen Franz Liszt und Carl Alexander, Herzog von Sachsen, ed. La Mara. Leipzig, 1909.

Briefwechsel zwischen Wagner und Liszt. Leipzig, 1919.

Correspondence de Liszt et de Madame d'Agoult. Paris, 1933. 2 vols.

Comtesse d'Agoult. Mémoires. Paris, 1927.

WORKS IN ENGLISH

Liszt's Letters, ed. Constance Bache. London, 1894.

Letters of Liszt and von Bülow, ed. Constance Bache. London, 1898.

Correspondence of Wagner and Liszt, ed. Francis Hueffer. London, 1888.

Letters of Franz Liszt to Marie zu Sayn-Wittgenstein, ed. Howard E. Hugo. Harvard, 1953.

Lina Ramann. Franz Liszt as Man and Artist. 1 vol. only translated. London, 1882.

James Huneker. Liszt. London, 1911.

A. Habets. Letters of Liszt and Borodin, ed. Rosa Newmarch. London, 1895.

Amy Fay. Music Study in Germany. London, 1893. (Dover Reprint, 1965)

Janka Wohl. François Liszt. London, 1887.

Life and Letters of Sir Charles Hallé. London, 1896.

Constance Bache. Brother Musicians. London, 1901.

Guy de Pourtalés. Franz Liszt, the Man of Love. London, 1927.

William Wallace. Liszt, Wagner and the Princess. London, 1927.

Sacheverell Sitwell. Liszt. London, 1934; revised edition, 1955. (Dover Reprint, 1966)

Ernest Newman. The Man Liszt. London, 1934. The Life of Richard Wagner. London, 1933.

Memoirs of Count Albert Apponyi. London, 1935.

Ralph Hill. Liszt. London, 1936.

Cecil Gray. Contingencies. London, 1947.

" Liszt " in Grove's Dictionary of Music, 5th edition. London, 1954.

Walter Beckett. Liszt. Master Musicians. London.

Bence Szabolcsi. The Twilight of Franz Liszt. Budapest.

New Hungarian Quarterly. Liszt-Bartók issue. Budapest, 1962.

PRINCIPAL WORKS IN OTHER LANGUAGES

Lina Ramann. Franz Liszt als Künstler und Mensch. Leipzig, 1880-94.

August Göllerich. Franz Liszt. Berlin, 1908.

Julius Kapp. Franz Liszt. Berlin, 1911.

Peter Raabe. Franz Liszt. Stuttgart, 1931.

Friedrich Schnapp. Verschollene Kompositionen Franz Liszts. In " Von Deutscher Tonkunst " (Festschrift für Peter Raabe). Peters, Leipzig, 1942.

INDEX OF WORKS MENTIONED

THIS AND THE following index apply only to the main text of the book (pp. 1-123). In this index the title of the work is immediately followed by its catalogue number, in italics; after the colon is the number of the page or pages on which it is referred to. For the sake of simplicity the fantasies are normally indexed under the titles of the works on which they are based, rather than under their correct titles (which will be found in the catalogue) —thus: Don Giovanni Fantasy; rather than: Réminiscences de Don Juan.

INDEX OF NAMES

INDEX OF NAMES

CATALOGUE OF DOVER BOOKS

Books Explaining Science and Mathematics

WHAT IS SCIENCE?, N. Campbell. The role of experiment and measurement, the function of mathematics, the nature of scientific laws, the difference between laws and theories, the limitations of science, and many similarly provocative topics are treated clearly and without technicalities by an eminent scientist. "Still an excellent introduction to scientific philosophy," H. Margenau in PHYSICS TODAY. "A first-rate primer . . . deserves a wide audience," SCIENTIFIC AMERICAN. 192pp. 5⅜ x 8.   S43 Paperbound **$1.25**

THE NATURE OF PHYSICAL THEORY, P. W. Bridgman. A Nobel Laureate's clear, non-technical lectures on difficulties and paradoxes connected with frontier research on the physical sciences. Concerned with such central concepts as thought, logic, mathematics, relativity, probability, wave mechanics, etc. he analyzes the contributions of such men as Newton, Einstein, Bohr, Heisenberg, and many others. "Lucid and entertaining . . . recommended to anyone who wants to get some insight into current philosophies of science," THE NEW PHILOSOPHY. Index. xi + 138pp. 5⅜ x 8.   S33 Paperbound **$1.25**

EXPERIMENT AND THEORY IN PHYSICS, Max Born. A Nobel Laureate examines the nature of experiment and theory in theoretical physics and analyzes the advances made by the great physicists of our day: Heisenberg, Einstein, Bohr, Planck, Dirac, and others. The actual process of creation is detailed step-by-step by one who participated. A fine examination of the scientific method at work. 44pp. 5⅜ x 8.   S308 Paperbound **75¢**

THE PSYCHOLOGY OF INVENTION IN THE MATHEMATICAL FIELD, J. Hadamard. The reports of such men as Descartes, Pascal, Einstein, Poincaré, and others are considered in this investigation of the method of idea-creation in mathematics and other sciences and the thinking process in general. How do ideas originate? What is the role of the unconscious? What is Poincaré's forgetting hypothesis? are some of the fascinating questions treated. A penetrating analysis of Einstein's thought processes concludes the book. xiii + 145pp. 5⅜ x 8.   T107 Paperbound **$1.25**

THE NATURE OF LIGHT AND COLOUR IN THE OPEN AIR, M. Minnaert. Why are shadows sometimes blue, sometimes green, or other colors depending on the light and surroundings? What causes mirages? Why do multiple suns and moons appear in the sky? Professor Minnaert explains these unusual phenomena and hundreds of others in simple, easy-to-understand terms based on optical laws and the properties of light and color. No mathematics is required but artists, scientists, students, and everyone fascinated by these "tricks" of nature will find thousands of useful and amazing pieces of information. Hundreds of observational experiments are suggested which require no special equipment. 200 illustrations; 42 photos. xvi + 362pp. 5⅜ x 8.   T196 Paperbound **$2.00**

THE UNIVERSE OF LIGHT, W. Bragg. Sir William Bragg, Nobel Laureate and great modern physicist, is also well known for his powers of clear exposition. Here he analyzes all aspects of light for the layman: lenses, reflection, refraction, the optics of vision, x-rays, the photoelectric effect, etc. He tells you what causes the color of spectra, rainbows, and soap bubbles, how magic mirrors work, and much more. Dozens of simple experiments are described. Preface. Index. 199 line drawings and photographs, including 2 full-page color plates. x + 283pp. 5⅜ x 8.   T538 Paperbound **$1.85**

SOAP-BUBBLES: THEIR COLOURS AND THE FORCES THAT MOULD THEM, C. V. Boys. For continuing popularity and validity as scientific primer, few books can match this volume of easily-followed experiments, explanations. Lucid exposition of complexities of liquid films, surface tension and related phenomena, bubbles' reaction to heat, motion, music, magnetic fields. Experiments with capillary attraction, soap bubbles on frames, composite bubbles, liquid cylinders and jets, bubbles other than soap, etc. Wonderful introduction to scientific method, natural laws that have many ramifications in areas of modern physics. Only complete edition in print. New Introduction by S. Z. Lewin, New York University. 83 illustrations; 1 full-page color plate. xii + 190pp. 5⅜ x 8½.   T542 Paperbound **95¢**

Puzzles, Mathematical Recreations

SYMBOLIC LOGIC and THE GAME OF LOGIC, Lewis Carroll. "Symbolic Logic" is not concerned with modern symbolic logic, but is instead a collection of over 380 problems posed with charm and imagination, using the syllogism, and a fascinating diagrammatic method of drawing conclusions. In "The Game of Logic" Carroll's whimsical imagination devises a logical game played with 2 diagrams and counters (included) to manipulate hundreds of tricky syllogisms. The final section, "Hit or Miss" is a lagniappe of 101 additional puzzles in the delightful Carroll manner. Until this reprint edition, both of these books were rarities costing up to $15 each. Symbolic Logic: Index. xxxi + 199pp. The Game of Logic: 96pp. 2 vols. bound as one. 5⅜ x 8. **T492 Paperbound $1.50**

PILLOW PROBLEMS and A TANGLED TALE, Lewis Carroll. One of the rarest of all Carroll's works, "Pillow Problems" contains 72 original math puzzles, all typically ingenious. Particularly fascinating are Carroll's answers which remain exactly as he thought them out, reflecting his actual mental process. The problems in "A Tangled Tale" are in story form, originally appearing as a monthly magazine serial. Carroll not only gives the solutions, but uses answers sent in by readers to discuss wrong approaches and misleading paths, and grades them for insight. Both of these books were rarities until this edition, "Pillow Problems" costing up to $25, and "A Tangled Tale" $15. Pillow Problems: Preface and Introduction by Lewis Carroll. xx + 109pp. A Tangled Tale: 6 illustrations. 152pp. Two vols. bound as one. 5⅜ x 8. **T493 Paperbound $1.50**

AMUSEMENTS IN MATHEMATICS, Henry Ernest Dudeney. The foremost British originator of mathematical puzzles is always intriguing, witty, and paradoxical in this classic, one of the largest collections of mathematical amusements. More than 430 puzzles, problems, and paradoxes. Mazes and games, problems on number manipulation, unicursal and other route problems, puzzles on measuring, weighing, packing, age, kinship, chessboards, joiners', crossing river, plane figure dissection, and many others. Solutions. More than 450 illustrations. vii + 258pp. 5⅜ x 8. **T473 Paperbound $1.25**

THE CANTERBURY PUZZLES, Henry Dudeney. Chaucer's pilgrims set one another problems in story form. Also Adventures of the Puzzle Club, the Strange Escape of the King's Jester, the Monks of Riddlewell, the Squire's Christmas Puzzle Party, and others. All puzzles are original, based on dissecting plane figures, arithmetic, algebra, elementary calculus and other branches of mathematics, and purely logical ingenuity. "The limit of ingenuity and intricacy," The Observer. Over 110 puzzles. Full Solutions. 150 illustrations. vii + 225pp. 5⅜ x 8. **T474 Paperbound $1.25**

MATHEMATICAL EXCURSIONS, H. A. Merrill. Even if you hardly remember your high school math, you'll enjoy the 90 stimulating problems contained in this book and you will come to understand a great many mathematical principles with surprisingly little effort. Many useful shortcuts and diversions not generally known are included: division by inspection, Russian peasant multiplication, memory systems for pi, building odd and even magic squares, square roots by geometry, dyadic systems, and many more. Solutions to difficult problems. 50 illustrations. 145pp. 5⅜ x 8. **T350 Paperbound $1.00**

MAGIC SQUARES AND CUBES, W. S. Andrews. Only book-length treatment in English, a thorough non-technical description and analysis. Here are nasik, overlapping, pandiagonal, serrated squares; magic circles, cubes, spheres, rhombuses. Try your hand at 4-dimensional magical figures! Much unusual folklore and tradition included. High school algebra is sufficient. 754 diagrams and illustrations. viii + 419pp. 5⅜ x 8. **T658 Paperbound $1.85**

CALIBAN'S PROBLEM BOOK: MATHEMATICAL, INFERENTIAL AND CRYPTOGRAPHIC PUZZLES, H. Phillips (Caliban), S. T. Shovelton, G. S. Marshall. 105 ingenious problems by the greatest living creator of puzzles based on logic and inference. Rigorous, modern, piquant; reflecting their author's unusual personality, these intermediate and advanced puzzles all involve the ability to reason clearly through complex situations; some call for mathematical knowledge, ranging from algebra to number theory. Solutions. xi + 180pp. 5⅜ x 8.
T736 Paperbound $1.25

MATHEMATICAL PUZZLES FOR BEGINNERS AND ENTHUSIASTS, G. Mott-Smith. 188 mathematical puzzles based on algebra, dissection of plane figures, permutations, and probability, that will test and improve your powers of inference and interpretation. The Odic Force, The Spider's Cousin, Ellipse Drawing, theory and strategy of card and board games like tit-tat-toe, go moku, salvo, and many others. 100 pages of detailed mathematical explanations. Appendix of primes, square roots, etc. 135 illustrations. 2nd revised edition. 248pp. 5⅜ x 8.
T198 Paperbound $1.00

MATHEMAGIC, MAGIC PUZZLES, AND GAMES WITH NUMBERS, R. V. Heath. More than 60 new puzzles and stunts based on the properties of numbers. Easy techniques for multiplying large numbers mentally, revealing hidden numbers magically, finding the date of any day in any year, and dozens more. Over 30 pages devoted to magic squares, triangles, cubes, circles, etc. Edited by J. S. Meyer. 76 illustrations. 128pp. 5⅜ x 8. **T110 Paperbound $1.00**

COMMON SPIDERS OF THE UNITED STATES, J. H. Emerton. Here is a nature hobby you can pursue right in your own cellar! Only non-technical, but thorough, reliable guide to spiders for the layman. Over 200 spiders from all parts of the country, arranged by scientific classification, are identified by shape and color, number of eyes, habitat and range, habits, etc. Full text, 501 line drawings and photographs, and valuable introduction explain webs, poisons, threads, capturing and preserving spiders, etc. Index. New synoptic key by S. W. Frost. xxiv + 225pp. 5⅜ x 8.   T223 Paperbound **$1.45**

THE LIFE STORY OF THE FISH: HIS MANNERS AND MORALS, Brian Curtis. A comprehensive, non-technical survey of just about everything worth knowing about fish. Written for the aquarist, the angler, and the layman with an inquisitive mind, the text covers such topics as evolution, external covering and protective coloration, physics and physiology of vision, maintenance of equilibrium, function of the lateral line canal for auditory and temperature senses, nervous system, function of the air bladder, reproductive system and methods—courtship, mating, spawning, care of young—and many more. Also sections on game fish, the problems of conservation and a fascinating chapter on fish curiosities. "Clear, simple language . . . excellent judgment in choice of subjects . . . delightful sense of humor," New York Times. Revised (1949) edition. Index. Bibliography of 72 items. 6 full-page photographic plates. xii + 284pp. 5⅜ x 8.   T929 Paperbound **$1.65**

BATS, Glover Morrill Allen. The most comprehensive study of bats as a life-form by the world's foremost authority. A thorough summary of just about everything known about this fascinating and mysterious flying mammal, including its unique location sense, hibernation and cycles, its habitats and distribution, its wing structure and flying habits, and its relationship to man in the long history of folklore and superstition. Written on a middle-level, the book can be profitably studied by a trained zoologist and thoroughly enjoyed by the layman. "An absorbing text with excellent illustrations. Bats should have more friends and fewer thoughtless detractors as a result of the publication of this volume," William Beebe, Books. Extensive bibliography. 57 photographs and illustrations. x + 368pp. 5⅜ x 8½.
T984 Paperbound **$2.00**

BIRDS AND THEIR ATTRIBUTES, Glover Morrill Allen. A fine general introduction to birds as living organisms, especially valuable because of emphasis on structure, physiology, habits, behavior. Discusses relationship of bird to man, early attempts at scientific ornithology, feathers and coloration, skeletal structure including bills, legs and feet, wings. Also food habits, evolution and present distribution, feeding and nest-building, still unsolved questions of migrations and location sense, many more similar topics. Final chapter on classification, nomenclature. A good popular-level summary for the biologist; a first-rate introduction for the layman. Reprint of 1925 edition. References and index. 51 illustrations. viii + 338pp. 5⅜ x 8½.   T957 Paperbound **$1.85**

LIFE HISTORIES OF NORTH AMERICAN BIRDS, Arthur Cleveland Bent. Bent's monumental series of books on North American birds, prepared and published under auspices of Smithsonian Institute, is the definitive coverage of the subject, the most-used single source of information. Now the entire set is to be made available by Dover in inexpensive editions. This encyclopedic collection of detailed, specific observations utilizes reports of hundreds of contemporary observers, writings of such naturalists as Audubon, Burroughs, William Brewster, as well as author's own extensive investigations. Contains literally everything known about life history of each bird considered: nesting, eggs, plumage, distribution and migration, voice, enemies, courtship, etc. These not over-technical works are musts for ornithologists, conservationists, amateur naturalists, anyone seriously interested in American birds.

BIRDS OF PREY. More than 100 subspecies of hawks, falcons, eagles, buzzards, condors and owls, from the common barn owl to the extinct caracara of Guadaloupe Island. 400 photographs. Two volume set. Index for each volume. Bibliographies of 403, 520 items. 197 full-page plates. Total of 907pp. 5⅜ x 8½.   Vol. I T931 Paperbound **$2.50**
                                                            Vol. II T932 Paperbound **$2.50**

WILD FOWL. Ducks, geese, swans, and tree ducks—73 different subspecies. Two volume set. Index for each volume. Bibliographies of 124, 144 items. 106 full-page plates. Total of 685pp. 5⅜ x 8½.   Vol. I T285 Paperbound **$2.50**
                                                            Vol. II T286 Paperbound **$2.50**

SHORE BIRDS. 81 varieties (sandpipers, woodcocks, plovers, snipes, phalaropes, curlews, oyster catchers, etc.). More than 200 photographs of eggs, nesting sites, adult and young of important species. Two volume set. Index for each volume. Bibliographies of 261, 188 items. 121 full-page plates. Total of 860pp. 5⅜ x 8½.   Vol. I T933 Paperbound **$2.35**
                                                            Vol. II T934 Paperbound **$2.35**

THE LIFE OF PASTEUR, R. Vallery-Radot. 13th edition of this definitive biography, cited in Encyclopaedia Britannica. Authoritative, scholarly, well-documented with contemporary quotes, observations; gives complete picture of Pasteur's personal life; especially thorough presentation of scientific activities with silkworms, fermentation, hydrophobia, inoculation, etc. Introduction by Sir William Osler. Index. 505pp. 5⅜ x 8.   T632 Paperbound **$2.00**

Nature, Biology

NATURE RECREATION: Group Guidance for the Out-of-doors, William Gould Vinal. Intended for both the uninitiated nature instructor and the education student on the college level, this complete "how-to" program surveys the entire area of nature education for the young. Philosophy of nature recreation; requirements, responsibilities, important information for group leaders; nature games; suggested group projects; conducting meetings and getting discussions started; etc. Scores of immediately applicable teaching aids, plus completely updated sources of information, pamphlets, field guides, recordings, etc. Bibliography. 74 photographs. + 310pp. 5⅜ x 8½.   T1015 Paperbound **$1.75**

HOW TO KNOW THE WILD FLOWERS, Mrs. William Starr Dana. Classic nature book that has introduced thousands to wonders of American wild flowers. Color-season principle of organization is easy to use, even by those with no botanical training, and the genial, refreshing discussions of history, folklore, uses of over 1,000 native and escape flowers, foliage plants are informative as well as fun to read. Over 170 full-page plates, collected from several editions, may be colored in to make permanent records of finds. Revised to conform with 1950 edition of Gray's Manual of Botany. xlii + 438pp. 5⅜ x 8½.   T332 Paperbound **$2.00**

HOW TO KNOW THE FERNS, F. T. Parsons. Ferns, among our most lovely native plants, are all too little known. This classic of nature lore will enable the layman to identify almost any American fern he may come across. After an introduction on the structure and life of ferns, the 57 most important ferns are fully pictured and described (arranged upon a simple identification key). Index of Latin and English names. 61 illustrations and 42 full-page plates. xiv + 215pp. 5⅜ x 8.   T740 Paperbound **$1.35**

MANUAL OF THE TREES OF NORTH AMERICA, Charles Sprague Sargent. Still unsurpassed as most comprehensive, reliable study of North American tree characteristics, precise locations and distribution. By dean of American dendrologists. Every tree native to U.S., Canada, Alaska, 185 genera, 717 species, described in detail—leaves, flowers, fruit, winterbuds, bark, wood, growth habits etc. plus discussion of varieties and local variants, immaturity variations. Over 100 keys, including unusual 11-page analytical key to genera, aid in identification. 783 clear illustrations of flowers, fruit, leaves. An unmatched permanent reference work for all nature lovers. Second enlarged (1926) edition. Synopsis of families. Analytical key to genera. Glossary of technical terms. Index. 783 illustrations, 1 map. Two volumes. Total of 982pp. 5⅜ x 8.   T277 Vol. I Paperbound **$2.25**
                                                      T278 Vol. II Paperbound **$2.25**
                                                                    The set **$4.50**

TREES OF THE EASTERN AND CENTRAL UNITED STATES AND CANADA, W. M. Harlow. A revised edition of a standard middle-level guide to native trees and important escapes. More than 140 trees are described in detail, and illustrated with more than 600 drawings and photographs. Supplementary keys will enable the careful reader to identify almost any tree he might encounter. xiii + 288pp. 5⅜ x 8.   T395 Paperbound **$1.35**

GUIDE TO SOUTHERN TREES, Ellwood S. Harrar and J. George Harrar. All the essential information about trees indigenous to the South, in an extremely handy format. Introductory essay on methods of tree classification and study, nomenclature, chief divisions of Southern trees, etc. Approximately 100 keys and synopses allow for swift, accurate identification of trees. Numerous excellent illustrations, non-technical text make this a useful book for teachers of biology or natural science, nature lovers, amateur naturalists. Revised 1962 edition. Index. Bibliography. Glossary of technical terms. 920 illustrations; 201 full-page plates. ix + 709pp. 4⅝ x 6⅜.   T945 Paperbound **$2.35**

FRUIT KEY AND TWIG KEY TO TREES AND SHRUBS, W. M. Harlow. Bound together in one volume for the first time, these handy and accurate keys to fruit and twig identification are the only guides of their sort with photographs (up to 3 times natural size). "Fruit Key": Key to over 120 different deciduous and evergreen fruits. 139 photographs and 11 line drawings. Synoptic summary of fruit types. Bibliography. 2 Indexes (common and scientific names). "Twig Key": Key to over 160 different twigs and buds. 173 photographs. Glossary of technical terms. Bibliography. 2 Indexes (common and scientific names). Two volumes bound as one. Total of xvii + 126pp. 5⅝ x 8⅜.   T511 Paperbound **$1.25**

INSECT LIFE AND INSECT NATURAL HISTORY, S. W. Frost. A work emphasizing habits, social life, and ecological relations of insects, rather than more academic aspects of classification and morphology. Prof. Frost's enthusiasm and knowledge are everywhere evident as he discusses insect associations and specialized habits like leaf-rolling, leaf-mining, and casemaking, the gall insects, the boring insects, aquatic insects, etc. He examines all sorts of matters not usually covered in general works, such as: insects as human food, insect music and musicians, insect response to electric and radio waves, use of insects in art and literature. The admirably executed purpose of this book, which covers the middle ground between elementary treatment and scholarly monographs, is to excite the reader to observe for himself. Over 700 illustrations. Extensive bibliography. x + 524pp. 5⅜ x 8.   T517 Paperbound **$2.45**

THE STORY OF X-RAYS FROM RONTGEN TO ISOTOPES, A. R. Bleich, M.D. This book, by a member of the American College of Radiology, gives the scientific explanation of x-rays, their applications in medicine, industry and art, and their danger (and that of atmospheric radiation) to the individual and the species. You learn how radiation therapy is applied against cancer, how x-rays diagnose heart disease and other ailments, how they are used to examine mummies for information on diseases of early societies, and industrial materials for hidden weaknesses. 54 illustrations show x-rays of flowers, bones, stomach, gears with flaws, etc. 1st publication. Index. xix + 186pp. 5⅜ x 8. T622 Paperbound **$1.35**

SPINNING TOPS AND GYROSCOPIC MOTION, John Perry. A classic elementary text of the dynamics of rotation — the behavior and use of rotating bodies such as gyroscopes and tops. In simple, everyday English you are shown how quasi-rigidity is induced in discs of paper, smoke rings, chains, etc., by rapid motions; why a gyrostat falls and why a top rises; precession; how the earth's motion affects climate; and many other phenomena. Appendix on practical use of gyroscopes. 62 figures. 128pp. 5⅜ x 8. T416 Paperbound **$1.00**

SNOW CRYSTALS, W. A. Bentley, M. J. Humphreys. For almost 50 years W. A. Bentley photographed snow flakes in his laboratory in Jericho, Vermont; in 1931 the American Meteorological Society gathered together the best of his work, some 2400 photographs of snow flakes, plus a few ice flowers, windowpane frosts, dew, frozen rain, and other ice formations. Pictures were selected for beauty and scientific value. A very valuable work to anyone in meteorology, cryology; most interesting to layman; extremely useful for artist who wants beautiful, crystalline designs. All copyright free. Unabridged reprint of 1931 edition. 2453 illustrations. 227pp. 8 x 10½. T287 Paperbound **$3.00**

A DOVER SCIENCE SAMPLER, edited by George Barkin. A collection of brief, non-technical passages from 44 Dover Books Explaining Science for the enjoyment of the science-minded browser. Includes work of Bertrand Russell, Poincaré, Laplace, Max Born, Galileo, Newton; material on physics, mathematics, metallurgy, anatomy, astronomy, chemistry, etc. You will be fascinated by Martin Gardner's analysis of the sincere pseudo-scientist, Moritz's account of Newton's absentmindedness, Bernard's examples of human vivisection, etc. Illustrations from the Diderot Pictorial Encyclopedia and De Re Metallica. 64 pages. **FREE**

THE STORY OF ATOMIC THEORY AND ATOMIC ENERGY, J. G. Feinberg. A broader approach to subject of nuclear energy and its cultural implications than any other similar source. Very readable, informal, completely non-technical text. Begins with first atomic theory, 600 B.C. and carries you through the work of Mendelejeff, Röntgen, Madame Curie, to Einstein's equation and the A-bomb. New chapter goes through thermonuclear fission, binding energy, other events up to 1959. Radioactive decay and radiation hazards, future benefits, work of Bohr, moderns, hundreds more topics. "Deserves special mention . . . not only authoritative but thoroughly popular in the best sense of the word," Saturday Review. Formerly, "The Atom Story." Expanded with new chapter. Three appendixes. Index. 34 illustrations. vii + 243pp. 5⅜ x 8. T625 Paperbound **$1.60**

THE STRANGE STORY OF THE QUANTUM, AN ACCOUNT FOR THE GENERAL READER OF THE GROWTH OF IDEAS UNDERLYING OUR PRESENT ATOMIC KNOWLEDGE, B. Hoffmann. Presents lucidly and expertly, with barest amount of mathematics, the problems and theories which led to modern quantum physics. Dr. Hoffmann begins with the closing years of the 19th century, when certain trifling discrepancies were noticed, and with illuminating analogies and examples takes you through the brilliant concepts of Planck, Einstein, Pauli, Broglie, Bohr, Schroedinger, Heisenberg, Dirac, Sommerfeld, Feynman, etc. This edition includes a new, long postscript carrying the story through 1958. "Of the books attempting an account of the history and contents of our modern atomic physics which have come to my attention, this is the best," H. Margenau, Yale University, in "American Journal of Physics." 32 tables and line illustrations. Index. 275pp. 5⅜ x 8. T518 Paperbound **$1.50**

SPACE AND TIME, E. Borel. Written by a versatile mathematician of world renown with his customary lucidity and precision, this introduction to relativity for the layman presents scores of examples, analogies, and illustrations that open up new ways of thinking about space and time. It covers abstract geometry and geographical maps, continuity and topology, the propagation of light, the special theory of relativity, the general theory of relativity, theoretical researches, and much more. Mathematical notes. 2 Indexes. 4 Appendices. 15 figures. xvi + 243pp. 5⅜ x 8. T592 Paperbound **$1.45**

FROM EUCLID TO EDDINGTON: A STUDY OF THE CONCEPTIONS OF THE EXTERNAL WORLD, Sir Edmund Whittaker. A foremost British scientist traces the development of theories of natural philosophy from the western rediscovery of Euclid to Eddington, Einstein, Dirac, etc. The inadequacy of classical physics is contrasted with present day attempts to understand the physical world through relativity, non-Euclidean geometry, space curvature, wave mechanics, etc. 5 major divisions of examination: Space; Time and Movement; the Concepts of Classical Physics; the Concepts of Quantum Mechanics; the Eddington Universe. 212pp. 5⅜ x 8. T491 Paperbound **$1.35**

THE HUMOROUS VERSE OF LEWIS CARROLL. Almost every poem Carroll ever wrote, the largest collection ever published, including much never published elsewhere: 150 parodies, burlesques, riddles, ballads, acrostics, etc., with 130 original illustrations by Tenniel, Carroll, and others. "Addicts will be grateful . . . there is nothing for the faithful to do but sit down and fall to the banquet," N. Y. Times. Index to first lines. xiv + 446pp. 5⅜ x 8.
T654 Paperbound **$2.00**

DIVERSIONS AND DIGRESSIONS OF LEWIS CARROLL. A major new treasure for Carroll fans! Rare privately published humor, fantasy, puzzles, and games by Carroll at his whimsical best, with a new vein of frank satire. Includes many new mathematical amusements and recreations, among them the fragmentary Part III of "Curiosa Mathematica." Contains "The Rectory Umbrella," "The New Belfry," "The Vision of the Three T's," and much more. New 32-page supplement of rare photographs taken by Carroll. x + 375pp. 5⅜ x 8.
T732 Paperbound **$2.00**

THE COMPLETE NONSENSE OF EDWARD LEAR. This is the only complete edition of this master of gentle madness available at a popular price. A BOOK OF NONSENSE, NONSENSE SONGS, MORE NONSENSE SONGS AND STORIES in their entirety with all the old favorites that have delighted children and adults for years. The Dong With A Luminous Nose, The Jumblies, The Owl and the Pussycat, and hundreds of other bits of wonderful nonsense. 214 limericks, 3 sets of Nonsense Botany, 5 Nonsense Alphabets, 546 drawings by Lear himself, and much more. 320pp. 5⅜ x 8.
T167 Paperbound **$1.00**

THE MELANCHOLY LUTE, The Humorous Verse of Franklin P. Adams ("FPA"). The author's own selection of light verse, drawn from thirty years of FPA's column, "The Conning Tower," syndicated all over the English-speaking world. Witty, perceptive, literate, these ninety-six poems range from parodies of other poets, Millay, Longfellow, Edgar Guest, Kipling, Masefield, etc., and free and hilarious translations of Horace and other Latin poets, to satiric comments on fabled American institutions—the New York Subways, preposterous ads, suburbanites, sensational journalism, etc. They reveal with vigor and clarity the humor, integrity and restraint of a wise and gentle American satirist. Introduction by Robert Hutchinson. vi + 122pp. 5⅜ x 8½.
T108 Paperbound **$1.00**

SINGULAR TRAVELS, CAMPAIGNS, AND ADVENTURES OF BARON MUNCHAUSEN, R. E. Raspe, with 90 illustrations by Gustave Doré. The first edition in over 150 years to reestablish the deeds of the Prince of Liars exactly as Raspe first recorded them in 1785—the genuine Baron Munchausen, one of the most popular personalities in English literature. Included also are the best of the many sequels, written by other hands. Introduction on Raspe by J. Carswell. Bibliography of early editions. xliv + 192pp. 5⅜ x 8.
T698 Paperbound **$1.00**

THE WIT AND HUMOR OF OSCAR WILDE, ed. by Alvin Redman. Wilde at his most brilliant, in 1000 epigrams exposing weaknesses and hypocrisies of "civilized" society. Divided into 49 categories—sin, wealth, women, America, etc.—to aid writers, speakers. Includes excerpts from his trials, books, plays, criticism. Formerly "The Epigrams of Oscar Wilde." Introduction by Vyvyan Holland, Wilde's only living son. Introductory essay by editor. 260pp. 5⅜ x 8.
T602 Paperbound **$1.00**

MAX AND MORITZ, Wilhelm Busch. Busch is one of the great humorists of all time, as well as the father of the modern comic strip. This volume, translated by H. A. Klein and other hands, contains the perennial favorite "Max and Moritz" (translated by C. T. Brooks), Plisch and Plum, Das Rabennest, Eispeter, and seven other whimsical, sardonic, jovial, diabolical cartoon and verse stories. Lively English translations parallel the original German. This work has delighted millions since it first appeared in the 19th century, and is guaranteed to please almost anyone. Edited by H. A. Klein, with an afterword. x + 205pp. 5⅝ x 8½.
T181 Paperbound **$1.15**

HYPOCRITICAL HELENA, Wilhelm Busch. A companion volume to "Max and Moritz," with the title piece (Die Fromme Helena) and 10 other highly amusing cartoon and verse stories, all newly translated by H. A. Klein and M. C. Klein: Adventure on New Year's Eve (Abenteuer in der Neujahrsnacht), Hangover on the Morning after New Year's Eve (Der Katzenjammer am Neujahrsmorgen), etc. English and German in parallel columns. Hours of pleasure, also a fine language aid. x + 205pp. 5⅝ x 8½.
T184 Paperbound **$1.00**

THE BEAR THAT WASN'T, Frank Tashlin. What does it mean? Is it simply delightful wry humor, or a charming story of a bear who wakes up in the midst of a factory, or a satire on Big Business, or an existential cartoon-story of the human condition, or a symbolization of the struggle between conformity and the individual? New York Herald Tribune said of the first edition: ". . . a fable for grownups that will be fun for children. Sit down with the book and get your own bearings." Long an underground favorite with readers of all ages and opinions. v + 51pp. Illustrated. 5⅜ x 8½.
T939 Paperbound **75¢**

RUTHLESS RHYMES FOR HEARTLESS HOMES and MORE RUTHLESS RHYMES FOR HEARTLESS HOMES, Harry Graham ("Col. D. Streamer"). Two volumes of Little Willy and 48 other poetic disasters. A bright, new reprint of oft-quoted, never forgotten, devastating humor by a precursor of today's "sick" joke school. For connoisseurs of wicked, wacky humor and all who delight in the comedy of manners. Original drawings are a perfect complement. 61 illustrations. Index. vi + 69pp. Two vols. bound as one. 5⅜ x 8½.
T930 Paperbound **75¢**

Say It language phrase books

These handy phrase books (128 to 196 pages each) make grammatical drills unnecessary for an elementary knowledge of a spoken foreign language. Covering most matters of travel and everyday life each volume contains:

Over 1000 phrases and sentences in immediately useful forms — foreign language plus English.

Modern usage designed for Americans. Specific phrases like, "Give me small change," and "Please call a taxi."

Simplified phonetic transcription you will be able to read at sight.

The only completely indexed phrase books on the market.

Covers scores of important situations: — Greetings, restaurants, sightseeing, useful expressions, etc.

These books are prepared by native linguists who are professors at Columbia, N.Y.U., Fordham and other great universities. Use them independently or with any other book or record course. They provide a supplementary living element that most other courses lack. Individual volumes in:

Russian 75¢	Italian 75¢	Spanish 75¢	German 75¢
Hebrew 75¢	Danish 75¢	Japanese 75¢	Swedish 75¢
Dutch 75¢	Esperanto 75¢	Modern Greek 75¢	Portuguese 75¢
Norwegian 75¢	Polish 75¢	French 75¢	Yiddish 75¢
Turkish 75¢		English for German-speaking people 75¢	
English for Italian-speaking people 75¢		English for Spanish-speaking people 75¢	

Large clear type. 128-196 pages each. 3½ x 5¼. Sturdy paper binding.

Listen and Learn language records

LISTEN & LEARN is the only language record course designed especially to meet your travel and everyday needs. It is available in separate sets for FRENCH, SPANISH, GERMAN, JAPANESE, RUSSIAN, MODERN GREEK, PORTUGUESE, ITALIAN and HEBREW, and each set contains three 33⅓ rpm long-playing records—1½ hours of recorded speech by eminent native speakers who are professors at Columbia, New York University, Queens College.

Check the following special features found only in LISTEN & LEARN:

● **Dual-language recording.** 812 selected phrases and sentences, over 3200 words, spoken first in English, then in their foreign language equivalents. A suitable pause follows each foreign phrase, allowing you time to repeat the expression. You learn by unconscious assimilation.

● **128 to 206-page manual** contains everything on the records, plus a simple phonetic pronunciation guide.

● **Indexed for convenience. The only set on the market** that is completely indexed. No more puzzling over where to find the phrase you need. Just look in the rear of the manual.

● **Practical.** No time wasted on material you can find in any grammar. LISTEN & LEARN covers central core material with phrase approach. Ideal for the person with limited learning time.

● **Living, modern expressions,** not found in other courses. Hygienic products, modern equipment, shopping—expressions used every day, like "nylon" and "air-conditioned."

● **Limited objective.** Everything you learn, no matter where you stop, is immediately useful. You have to finish other courses, wade through grammar and vocabulary drill, before they help you.

● **High-fidelity recording.** LISTEN & LEARN records equal in clarity and surface-silence any record on the market costing up to $6.

"Excellent . . . the spoken records . . . impress me as being among the very best on the market," **Prof. Mario Pei,** Dept. of Romance Languages, Columbia University. "Inexpensive and well-done . . . it would make an ideal present," CHICAGO SUNDAY TRIBUNE. "More genuinely helpful than anything of its kind which I have previously encountered," **Sidney Clark,** well-known author of "ALL THE BEST" travel books.

UNCONDITIONAL GUARANTEE. Try LISTEN & LEARN, then return it within 10 days for full refund if you are not satisfied.

Each set contains three twelve-inch 33⅓ records, manual, and album.

SPANISH	the set **$5.95**	GERMAN	the set **$5.95**
FRENCH	the set **$5.95**	ITALIAN	the set **$5.95**
RUSSIAN	the set **$5.95**	JAPANESE	the set **$5.95**
PORTUGUESE	the set **$5.95**	MODERN GREEK	the set **$5.95**
MODERN HEBREW	the set **$5.95**		

Americana

THE EYES OF DISCOVERY, J. Bakeless. A vivid reconstruction of how unspoiled America appeared to the first white men. Authentic and enlightening accounts of Hudson's landing in New York, Coronado's trek through the Southwest; scores of explorers, settlers, trappers, soldiers. America's pristine flora, fauna, and Indians in every region and state in fresh and unusual new aspects. "A fascinating view of what the land was like before the first highway went through," Time. 68 contemporary illustrations, 39 newly added in this edition. Index. Bibliography. x + 500pp. 5⅜ x 8.   T761 Paperbound **$2.00**

AUDUBON AND HIS JOURNALS, J. J. Audubon. A collection of fascinating accounts of Europe and America in the early 1800's through Audubon's own eyes. Includes the Missouri River Journals —an eventful trip through America's untouched heartland, the Labrador Journals, the European Journals, the famous "Episodes", and other rare Audubon material, including the descriptive chapters from the original letterpress edition of the "Ornithological Studies", omitted in all later editions. Indispensable for ornithologists, naturalists, and all lovers of Americana and adventure. 70-page biography by Audubon's granddaughter. 38 illustrations. Index. Total of 1106pp. 5⅜ x 8.   T675 Vol I Paperbound **$2.25**
                                                                                     T676 Vol II Paperbound **$2.25**
                                                                                     The set **$4.50**

TRAVELS OF WILLIAM BARTRAM, edited by Mark Van Doren. The first inexpensive illustrated edition of one of the 18th century's most delightful books is an excellent source of first-hand material on American geography, anthropology, and natural history. Many descriptions of early Indian tribes are our only source of information on them prior to the infiltration of the white man. "The mind of a scientist with the soul of a poet," John Livingston Lowes. 13 original illustrations and maps. Edited with an introduction by Mark Van Doren. 448pp. 5⅜ x 8.
                                                                                     T13 Paperbound **$2.00**

GARRETS AND PRETENDERS: A HISTORY OF BOHEMIANISM IN AMERICA, A. Parry. The colorful and fantastic history of American Bohemianism from Poe to Kerouac. This is the only complete record of hoboes, cranks, starving poets, and suicides. Here are Pfaff, Whitman, Crane, Bierce, Pound, and many others. New chapters by the author and by H. T. Moore bring this thorough and well-documented history down to the Beatniks. "An excellent account," N. Y. Times. Scores of cartoons, drawings, and caricatures. Bibliography. Index. xxviii + 421pp. 5⅝ x 8⅜.   T708 Paperbound **$1.95**

THE EXPLORATION OF THE COLORADO RIVER AND ITS CANYONS, J. W. Powell. The thrilling first-hand account of the expedition that filled in the last white space on the map of the United States. Rapids, famine, hostile Indians, and mutiny are among the perils encountered as the unknown Colorado Valley reveals its secrets. This is the only uncut version of Major Powell's classic of exploration that has been printed in the last 60 years. Includes later reflections and subsequent expedition. 250 illustrations, new map. 400pp. 5⅝ x 8⅜.
                                                                                     T94 Paperbound **$2.25**

THE JOURNAL OF HENRY D. THOREAU, Edited by Bradford Torrey and Francis H. Allen. Henry Thoreau is not only one of the most important figures in American literature and social thought; his voluminous journals (from which his books emerged as selections and crystalliza-tions) constitute both the longest, most sensitive record of personal internal development and a most penetrating description of a historical moment in American culture. This present set, which was first issued in fourteen volumes, contains Thoreau's entire journals from 1837 to 1862, with the exception of the lost years which were found only recently. We are reissuing it, complete and unabridged, with a new introduction by Walter Harding, Secretary of the Thoreau Society. Fourteen volumes reissued in two volumes. Foreword by Henry Seidel Canby. Total of 1888pp. 8⅜ x 12¼.   T312-3 Two volume set, Clothbound **$20.00**

GAMES AND SONGS OF AMERICAN CHILDREN, collected by William Wells Newell. A remarkable collection of 190 games with songs that accompany many of them; cross references to show similarities, differences among them; variations; musical notation for 38 songs. Textual dis-cussions show relations with folk-drama and other aspects of folk tradition. Grouped into categories for ready comparative study: Love-games, histories, playing at work, human life, bird and beast, mythology, guessing-games, etc. New introduction covers relations of songs and dances to timeless heritage of folklore, biographical sketch of Newell, other pertinent data. A good source of inspiration for those in charge of groups of children and a valuable reference for anthropologists, sociologists, psychiatrists. Introduction by Carl Withers. New indexes of first lines, games. 5⅜ x 8½. xii + 242pp.   T354 Paperbound **$1.75**

Art, History of Art, Antiques, Graphic Arts, Handcrafts

ART STUDENTS' ANATOMY, E. J. Farris. Outstanding art anatomy that uses chiefly living objects for its illustrations. 71 photos of undraped men, women, children are accompanied by carefully labeled matching sketches to illustrate the skeletal system, articulations and movements, bony landmarks, the muscular system, skin, fasciae, fat, etc. 9 x-ray photos show movement of joints. Undraped models are shown in such actions as serving in tennis, drawing a bow in archery, playing football, dancing, preparing to spring and to dive. Also discussed and illustrated are proportions, age and sex differences, the anatomy of the smile, etc. 8 plates by the great early 18th century anatomic illustrator Siegfried Albinus are also included. Glossary. 158 figures, 7 in color. x + 159pp. 5⅝ x 8⅜. **T744 Paperbound $1.50**

AN ATLAS OF ANATOMY FOR ARTISTS, F Schider. A new 3rd edition of this standard text enlarged by 52 new illustrations of hands, anatomical studies by Cloquet, and expressive life studies of the body by Barcsay. 189 clear, detailed plates offer you precise information of impeccable accuracy. 29 plates show all aspects of the skeleton, with closeups of special areas, while 54 full-page plates, mostly in two colors, give human musculature as seen from four different points of view, with cutaways for important portions of the body. 14 full-page plates provide photographs of hand forms, eyelids, female breasts, and indicate the location of muscles upon models. 59 additional plates show how great artists of the past utilized human anatomy. They reproduce sketches and finished work by such artists as Michelangelo, Leonardo da Vinci, Goya, and 15 others. This is a lifetime reference work which will be one of the most important books in any artist's library. "The standard reference tool," AMERICAN LIBRARY ASSOCIATION. "Excellent," AMERICAN ARTIST. Third enlarged edition. 189 plates, 647 illustrations. xxvi + 192pp. 7⅞ x 10⅝. **T241 Clothbound $6.00**

AN ATLAS OF ANIMAL ANATOMY FOR ARTISTS, W. Ellenberger, H. Baum, H. Dittrich. The largest, richest animal anatomy for artists available in English. 99 detailed anatomical plates of such animals as the horse, dog, cat, lion, deer, seal, kangaroo, flying squirrel, cow, bull, goat, monkey, hare, and bat. Surface features are clearly indicated, while progressive beneath-the-skin pictures show musculature, tendons, and bone structure. Rest and action are exhibited in terms of musculature and skeletal structure and detailed cross-sections are given for heads and important features. The animals chosen are representative of specific families so that a study of these anatomies will provide knowledge of hundreds of related species. "Highly recommended as one of the very few books on the subject worthy of being used as an authoritative guide," DESIGN. "Gives a fundamental knowledge," AMERICAN ARTIST. Second revised, enlarged edition with new plates from Cuvier, Stubbs, etc. 288 illustrations. 153pp. 11⅜ x 9. **T82 Clothbound $6.00**

THE HUMAN FIGURE IN MOTION, Eadweard Muybridge. The largest selection in print of Muybridge's famous high-speed action photos of the human figure in motion. 4789 photographs illustrate 162 different actions: men, women, children—mostly undraped—are shown walking, running, carrying various objects, sitting, lying down, climbing, throwing, arising, and performing over 150 other actions. Some actions are shown in as many as 150 photographs each. All in all there are more than 500 action strips in this enormous volume, series shots taken at shutter speeds of as high as 1/6000th of a second! These are not posed shots, but true stopped motion. They show bone and muscle in situations that the human eye is not fast enough to capture. Earlier, smaller editions of these prints have brought $40 and more on the out-of-print market. "A must for artists," ART IN FOCUS. "An unparalleled dictionary of action for all artists," AMERICAN ARTIST. 390 full-page plates, with 4789 photographs. Printed on heavy glossy stock. Reinforced binding with headbands. xxi + 390pp. 7⅞ x 10⅝. **T204 Clothbound $10.00**

ANIMALS IN MOTION, Eadweard Muybridge. This is the largest collection of animal action photos in print. 34 different animals (horses, mules, oxen, goats, camels, pigs, cats, guanacos, lions, gnus, deer, monkeys, eagles—and 21 others) in 132 characteristic actions. The horse alone is shown in more than 40 different actions. All 3919 photographs are taken in series at speeds up to 1/6000th of a second. The secrets of leg motion, spinal patterns, head movements, strains and contortions shown nowhere else are captured. You will see exactly how a lion sets his foot down; how an elephant's knees are like a human's—and how they differ; the position of a kangaroo's legs in mid-leap; how an ostrich's head bobs; details of the flight of birds—and thousands of facets of motion only the fastest cameras can catch. Photographed from domestic animals and animals in the Philadelphia zoo, it contains neither semiposed artificial shots nor distorted telephoto shots taken under adverse conditions. Artists, biologists, decorators, cartoonists, will find this book indispensable for understanding animals in motion. "A really marvelous series of plates," NATURE (London). "The dry plate's most spectacular early use was by Eadweard Muybridge," LIFE. 3919 photographs; 380 full pages of plates. 440pp. Printed on heavy glossy paper. Deluxe binding with headbands. 7⅞ x 10⅝. **T203 Clothbound $10.00**

THE AUTOBIOGRAPHY OF AN IDEA, Louis Sullivan. The pioneer architect whom Frank Lloyd Wright called "the master" reveals an acute sensitivity to social forces and values in this passionately honest account. He records the crystallization of his opinions and theories, the growth of his organic theory of architecture that still influences American designers and architects, contemporary ideas, etc. This volume contains the first appearance of 34 full-page plates of his finest architecture. Unabridged reissue of 1924 edition. New introduction by R. M. Line. Index. xiv + 335pp. 5⅜ x 8. **T281 Paperbound $2.00**

THE DRAWINGS OF HEINRICH KLEY. The first uncut republication of both of Kley's devastating sketchbooks, which first appeared in pre-World War I Germany. One of the greatest cartoonists and social satirists of modern times, his exuberant and iconoclastic fantasy and his extra-ordinary technique place him in the great tradition of Bosch, Breughel, and Goya, while his subject matter has all the immediacy and tension of our century. 200 drawings. viii + 128pp. 7¾ x 10¾. **T24 Paperbound $1.85**

MORE DRAWINGS BY HEINRICH KLEY. All the sketches from Leut' Und Viecher (1912) and Sammel-Album (1923) not included in the previous Dover edition of Drawings. More of the bizarre, mercilessly iconoclastic sketches that shocked and amused on their original publica-tion. Nothing was too sacred, no one too eminent for satirization by this imaginative, in-dividual and accomplished master cartoonist. A total of 158 illustrations. Iv + 104pp. 7¾ x 10¾. **T41 Paperbound $1.85**

PINE FURNITURE OF EARLY NEW ENGLAND, R. H. Kettell. A rich understanding of one of America's most original folk arts that collectors of antiques, interior decorators, craftsmen, woodworkers, and everyone interested in American history and art will find fascinating and immensely useful. 413 illustrations of more than 300 chairs, benches, racks, beds, cupboards, mirrors, shelves, tables, and other furniture will show all the simple beauty and character of early New England furniture. 55 detailed drawings carefully analyze outstanding pieces. "With its rich store of illustrations, this book emphasizes the individuality and varied design of early American pine furniture. It should be welcomed," ANTIQUES. 413 illustrations and 55 working drawings. 475. 8 x 10¾. **T145 Clothbound $10.00**

THE HUMAN FIGURE, J. H. Vanderpoel. Every important artistic element of the human figure is pointed out in minutely detailed word descriptions in this classic text and illustrated as well in 430 pencil and charcoal drawings. Thus the text of this book directs your attention to all the characteristic features and subtle differences of the male and female (adults, children, and aged persons), as though a master artist were telling you what to look for at each stage. 2nd edition, revised and enlarged by George Bridgman. Foreword. 430 illustrations. 143pp. 6⅛ x 9¼. **T432 Paperbound $1.50**

LETTERING AND ALPHABETS, J. A. Cavanagh. This unabridged reissue of LETTERING offers a full discussion, analysis, illustration of 89 basic hand lettering styles — styles derived from Caslons, Bodonis, Garamonds, Gothic, Black Letter, Oriental, and many others. Upper and lower cases, numerals and common signs pictured. Hundreds of technical hints on make-up, construction, artistic validity, strokes, pens, brushes, white areas, etc. May be reproduced without permission! 89 complete alphabets; 72 lettered specimens. 121pp. 9¾ x 8. **T53 Paperbound $1.35**

STICKS AND STONES, Lewis Mumford. A survey of the forces that have conditioned American architecture and altered its forms. The author discusses the medieval tradition in early New England villages; the Renaissance influence which developed with the rise of the merchant class; the classical influence of Jefferson's time; the "Mechanicsvilles" of Poe's generation; the Brown Decades; the philosophy of the Imperial facade; and finally the modern machine age. "A truly remarkable book," SAT. REV. OF LITERATURE. 2nd revised edition. 21 illustra-tions. xvii + 228pp. 5⅜ x 8. **T202 Paperbound $1.75**

THE STANDARD BOOK OF QUILT MAKING AND COLLECTING, Marguerite Ickis. A complete easy-to-follow guide with all the information you need to make beautiful, useful quilts. How to plan, design, cut, sew, appliqué, avoid sewing problems, use rag bag, make borders, tuft, every other aspect. Over 100 traditional quilts shown, including over 40 full-size patterns. At-home hobby for fun, profit. Index. 483 illus. 1 color plate. 287pp. 6¾ x 9½. **T582 Paperbound $2.00**

THE BOOK OF SIGNS, Rudolf Koch. Formerly $20 to $25 on the out-of-print market, now only $1.00 in this unabridged new edition! 493 symbols from ancient manuscripts, medieval cathe-drals, coins, catacombs, pottery, etc. Crosses, monograms of Roman emperors, astrological, chemical, botanical, runes, housemarks, and 7 other categories. Invaluable for handicraft workers, illustrators, scholars, etc., this material may be reproduced without permission. 493 illustrations by Fritz Kredel. 104pp. 6½ x 9¼. **T162 Paperbound $1.00**

PRIMITIVE ART, Franz Boas. This authoritative and exhaustive work by a great American anthropologist covers the entire gamut of primitive art. Pottery, leatherwork, metal work, stone work, wood, basketry, are treated in detail. Theories of primitive art, historical depth in art history, technical virtuosity, unconscious levels of patterning, symbolism, styles, litera-ture, music, dance, etc. A must book for the interested layman, the anthropologist, artist, handicrafter (hundreds of unusual motifs), and the historian. Over 900 illustrations (50 ceramic vessels, 12 totem poles, etc.). 376pp. 5⅜ x 8. **T25 Paperbound $2.00**

Fiction

FLATLAND, E. A. Abbott. A science-fiction classic of life in a 2-dimensional world that is also a first-rate introduction to such aspects of modern science as relativity and hyperspace. Political, moral, satirical, and humorous overtones have made FLATLAND fascinating reading for thousands. 7th edition. New introduction by Banesh Hoffmann. 16 illustrations. 128pp. 5⅜ x 8. **T1 Paperbound $1.00**

THE WONDERFUL WIZARD OF OZ, L. F. Baum. Only edition in print with all the original W. W. Denslow illustrations in full color—as much a part of "The Wizard" as Tenniel's drawings are of "Alice in Wonderland." "The Wizard" is still America's best-loved fairy tale, in which, as the author expresses it, "The wonderment and joy are retained and the heartaches and nightmares left out." Now today's young readers can enjoy every word and wonderful picture of the original book. New introduction by Martin Gardner. A Baum bibliography. 23 full-page color plates. viii + 268pp. 5⅜ x 8. **T691 Paperbound $1.50**

THE MARVELOUS LAND OF OZ, L. F. Baum. This is the equally enchanting sequel to the "Wizard," continuing the adventures of the Scarecrow and the Tin Woodman. The hero this time is a little boy named Tip, and all the delightful Oz magic is still present. This is the Oz book with the Animated Saw-Horse, the Woggle-Bug, and Jack Pumpkinhead. All the original John R. Neill illustrations, 10 in full color. 287 pp. 5⅜ x 8. **T692 Paperbound $1.50**

28 SCIENCE FICTION STORIES OF H. G. WELLS. Two full unabridged novels, MEN LIKE GODS and STAR BEGOTTEN, plus 26 short stories by the master science-fiction writer of all time! Stories of space, time, invention, exploration, future adventure—an indispensable part of the library of everyone interested in science and adventure. PARTIAL CONTENTS: Men Like Gods, The Country of the Blind, In the Abyss, The Crystal Egg, The Man Who Could Work Miracles, A Story of the Days to Come, The Valley of Spiders, and 21 more! 928pp. 5⅜ x 8. **T265 Clothbound $4.50**

THREE MARTIAN NOVELS, Edgar Rice Burroughs. Contains: Thuvia, Maid of Mars; The Chessmen of Mars; and The Master Mind of Mars. High adventure set in an imaginative and intricate conception of the Red Planet. Mars is peopled with an intelligent, heroic human race which lives in densely populated cities and with fierce barbarians who inhabit dead sea bottoms. Other exciting creatures abound amidst an inventive framework of Martian history and geography. Complete unabridged reprintings of the first edition. 16 illustrations by J. Allen St. John. vi + 499pp. 5⅜ x 8½. **T39 Paperbound $1.85**

SEVEN SCIENCE FICTION NOVELS, H. G. Wells. Full unabridged texts of 7 science-fiction novels of the master. Ranging from biology, physics, chemistry, astronomy to sociology and other studies, Mr. Wells extrapolates whole worlds of strange and intriguing character. "One will have to go far to match this for entertainment, excitement, and sheer pleasure . . . ," NEW YORK TIMES. Contents: The Time Machine, The Island of Dr. Moreau, First Men in the Moon, The Invisible Man, The War of the Worlds, The Food of the Gods, In the Days of the Comet. 1015pp. 5⅜ x 8. **T264 Clothbound $4.50**

THE LAND THAT TIME FORGOT and THE MOON MAID, Edgar Rice Burroughs. In the opinion of many, Burroughs' best work. The first concerns a strange island where evolution is individual rather than phylogenetic. Speechless anthropoids develop into intelligent human beings within a single generation. The second projects the reader far into the future and describes the first voyage to the Moon (in the year 2025), the conquest of the Earth by the Moon, and years of violence and adventure as the enslaved Earthmen try to regain possession of their planet. "An imaginative tour de force that keeps the reader keyed up and expectant," NEW YORK TIMES. Complete, unabridged text of the original two novels (three parts in each). 5 illustrations by J. Allen St. John. vi + 552pp. 5⅜ x 8½.
T1020 Clothbound $3.75
T358 Paperbound $2.00

3 ADVENTURE NOVELS by H. Rider Haggard. Complete texts of "She," "King Solomon's Mines," "Allan Quatermain." Qualities of discovery; desire for immortality; search for primitive, for what is unadorned by civilization, have kept these novels of African adventure exciting, alive to readers from R. L. Stevenson to George Orwell. 636pp. 5⅜ x 8. **T584 Paperbound $2.00**

A PRINCESS OF MARS and A FIGHTING MAN OF MARS: TWO MARTIAN NOVELS BY EDGAR RICE BURROUGHS. "Princess of Mars" is the very first of the great Martian novels written by Burroughs, and it is probably the best of them all; it set the pattern for all of his later fantasy novels and contains a thrilling cast of strange peoples and creatures and the formula of Olympian heroism amidst ever-fluctuating fortunes which Burroughs carries off so successfully. "Fighting Man" returns to the same scenes and cities—many years later. A mad scientist, a degenerate dictator, and an indomitable defender of the right clash—with the fate of the Red Planet at stake! Complete, unabridged reprinting of original editions. Illustrations by F. E. Schoonover and Hugh Hutton. v + 356pp. 5⅜ x 8½. **T1140 Paperbound $1.75**

Music

A GENERAL HISTORY OF MUSIC, Charles Burney. A detailed coverage of music from the Greeks up to 1789, with full information on all types of music: sacred and secular, vocal and instrumental, operatic and symphonic. Theory, notation, forms, instruments, innovators, composers, performers, typical and important works, and much more in an easy, entertaining style. Burney covered much of Europe and spoke with hundreds of authorities and composers so that this work is more than a compilation of records . . . it is a living work of careful and first-hand scholarship. Its account of thoroughbass (18th century) Italian music is probably still the best introduction on the subject. A recent NEW YORK TIMES review said, "Surprisingly few of Burney's statements have been invalidated by modern research . . . still of great value." Edited and corrected by Frank Mercer. 35 figures. Indices. 1915pp. 5⅜ x 8. 2 volumes.　　　　　　　　　　　　　　　　　　**T36 The Set, Clothbound $12.50**

A DICTIONARY OF HYMNOLOGY, John Julian. This exhaustive and scholarly work has become known as an invaluable source of hundreds of thousands of important and often difficult to obtain facts on the history and use of hymns in the western world. Everyone interested in hymns will be fascinated by the accounts of famous hymns and hymn writers and amazed by the amount of practical information he will find. More than 30,000 entries on individual hymns, giving authorship, date and circumstances of composition, publication, textual variations, translations, denominational and ritual usage, etc. Biographies of more than 9,000 hymn writers, and essays on important topics such as Christmas carols and children's hymns, and much other unusual and valuable information. A 200 page double-columned index of first lines — the largest in print. Total of 1786 pages in two reinforced clothbound volumes. 6¼ x 9¼.
The set, T333 Clothbound $17.50

MUSIC IN MEDIEVAL BRITAIN, F. Ll. Harrison. The most thorough, up-to-date, and accurate treatment of the subject ever published, beautifully illustrated. Complete account of institutions and choirs; carols, masses, and motets; liturgy and plainsong; and polyphonic music from the Norman Conquest to the Reformation. Discusses the various schools of music and their reciprocal influences; the origin and development of new ritual forms; development and use of instruments; and new evidence on many problems of the period. Reproductions of scores, over 200 excerpts from medieval melodies. Rules of harmony and dissonance; influence of Continental styles; great composers (Dunstable, Cornysh, Fairfax, etc.); and much more. Register and index of more than 400 musicians. Index of titles. General Index. 225-item bibliography. 6 Appendices. xix + 491pp. 5⅝ x 8¾.　　　　**T705 Clothbound $10.00**

THE MUSIC OF SPAIN, Gilbert Chase. Only book in English to give concise, comprehensive account of Iberian music; new Chapter covers music since 1941. Victoria, Albéniz, Cabezón, Pedrell, Turina, hundreds of other composers; popular and folk music; the Gypsies; the guitar; dance, theatre, opera, with only extensive discussion in English of the Zarzuela; virtuosi such as Casals; much more. "Distinguished . . . readable," Saturday Review. 400-item bibliography. Index. 27 photos. 383pp. 5⅜ x 8.　　　　**T549 Paperbound $2.00**

ON STUDYING SINGING, Sergius Kagen. An intelligent method of voice-training, which leads you around pitfalls that waste your time, money, and effort. Exposes rigid, mechanical systems, baseless theories, deleterious exercises. "Logical, clear, convincing . . . dead right," Virgil Thomson, N.Y. Herald Tribune. "I recommend this volume highly," Maggie Teyte, Saturday Review. 119pp. 5⅜ x 8.　　　　　　　　　　**T622 Paperbound $1.35**

Prices subject to change without notice.

Dover publishes books on art, music, philosophy, literature, languages, history, social sciences, psychology, handcrafts, orientalia, puzzles and entertainments, chess, pets and gardens, books explaining science, intermediate and higher mathematics, mathematical physics, engineering, biological sciences, earth sciences, classics of science, etc. Write to:

Dept. catrr.
Dover Publications, Inc.
180 Varick Street, N.Y. 14, N.Y.